STUDENT SOLUTIONS MANUAL
Laurel Technical Services

Sixth Edition
FINITE MATHEMATICS
& Its Applications

GOLDSTEIN ▪ SCHNEIDER ▪ SIEGEL

PRENTICE HALL Upper Saddle River, NJ 07458

Supplements Editor: *April Thrower*
Production Editor: *Mindy DePalma*
Special Projects Manager: *Barbara A. Murray*
Supplement Cover Manager: *Paul Gourhan*
Production Coordinator: *Alan Fischer*

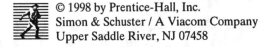

© 1998 by Prentice-Hall, Inc.
Simon & Schuster / A Viacom Company
Upper Saddle River, NJ 07458

Printed in the United States of America

10 9 8 7 6 5 4

ISBN 0-13-747684-1

Prentice-Hall International (UK) Limited, *London*
Prentice-Hall of Australia Pty. Limited, *Sydney*
Prentice-Hall Canada, Inc., *Toronto*
Prentice-Hall Hispanoamericana, S.A., *Mexico*
Prentice-Hall of India Private Limited, *New Delhi*
Prentice-Hall of Japan, Inc., *Tokyo*
Simon & Schuster Asia Pte. Ltd., *Singapore*
Editora Prentice-Hall do Brasil, Ltda., *Rio de Janeiro*

Table of Contents

Explorations in Finite Mathematics Software

The software packaged with this booklet enhances the understanding and visualization of many of the topics covered in the textbook. The self-documented software consists of two programs called FINITE1 and FINITE2.

Part 1 (Execute by entering FINITE1)

Part 1 covers finite math topics involving linear equations, linear inequalities, and matrices. In particular, it contains routines to perform Gaussian elimination, matrix operations, matrix inversion, simplex method, graphical solution of linear programming problems, the method of least-squares, and regular and absorbing Markov chains. The software follows the conventions of the text.

Part 2 (Execute by entering FINITE2)

part 2 contains routines for Venn diagrams, Venn diagram counting problems, Venn diagram probability problems, computation of combinations, permutations, and factorials, statistical analysis of data, Galton board, binomial distribution, areas under normal curve, simple interest, compound interest, loan analysis, annuity analysis, finance tables, difference equations, and truth tables.

Requirements

The software runs on *any* IBM compatible computer having at least 384 K of memory and a graphics adapter, that is CGA, EGA, MCGA, VGA, or Hercules.

Matrices On Diskette

About fifty matrices appearing in examples from the text have been saved on diskette with either suggestive names or names of the form CxSxEx. For instance, the matrix appearing in Chapter 8, Section 3, Example 7 has the name C8S3E7. The list of saved matrices appropriate to the current routine is displayed on the screen whenever you request "Load a matrix saved on disk."

To Invoke Explorations in Finite Mathematics from DOS

1. Place the diskette in a drive, say drive A.

2. Type A: and press the Enter key.

3. Type FINITE1 or FINITE2 and press the Enter key.
   (*Note*: If a Hercules adapter is used, the program MSHERC.COM must be run before FINITE1 or FINITE2 is entered.)

To Invoke Explorations in Finite Mathematics from Windows 3.1

1. Place the diskette in a drive, say drive A.

2. From Program Manager, double-click on the DOS icon in the Main program group. Or, exit windows.

3. Type A: and press the Enter key.

4. Type FINITE1 or FINITE2 and press the Enter key.

To Invoke Explorations in Finite Mathematics from Windows 95 (or later version)

1. Place the diskette in a drive, say drive A.

2. Click the Start button, point to Programs, and click MS-DOS Prompt.

3. If the DOS window does not fill the screen, hold down the Alt key and press the Enter to enlarge the window.

4. Type A: and press the Enter key.

5. Type FINITE1 or FINITE2 and press the Enter key.

Chapter 1

Exercises 1

1. Right 2, up 3

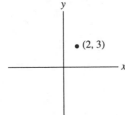

3. Down 2

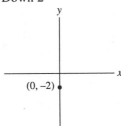

5. Left 2, up 1

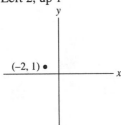

7. Left 20, up 40

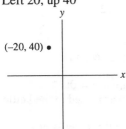

9. Right 7, down 5: $(7, -5)$, or (e)

11. $m = 5, b = 8$

13. $y = 0x + 3; m = 0, b = 3$

15. $14x + 7y = 21$
$7y = -14x + 21$
$y = -2x + 3$

17. $3x = 5$
$x = \dfrac{5}{3}$

19. $0 = -4x + 8$
$4x = 8$
$x = 2$
x-intercept: $(2, 0)$
$y = -4(0) + 8$
$y = 8$
y-intercept: $(0, 8)$

21. When $y = 0, x = 7$
x-intercept: $(7, 0)$
$0 = 7$
no solution
y-intercept: none

23. $0 = \dfrac{1}{3}x - 1$
$x = 3$
x-intercept: $(3, 0)$
$y = \dfrac{1}{3}(0) - 1$
$y = -1$
y-intercept: $(0, -1)$

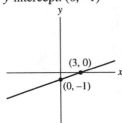

25. $0 = \dfrac{5}{2}$
no solution
x-intercept: none

When $x = 0$, $y = \dfrac{5}{2}$

y-intercept: $\left(0, \dfrac{5}{2}\right)$

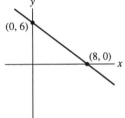

27. $3x + 4(0) = 24$
$x = 8$
x-intercept: $(8, 0)$
$3(0) + 4y = 24$
$y = 6$
y-intercept: $(0, 6)$

29. $2x + 3y = 6$
$3y = -2x + 6$
$y = -\dfrac{2}{3}x + 2$

  a. $4x + 6y = 12$
$6y = -4x + 12$
$y = -\dfrac{2}{3}x + 2$
Yes

  b.   Yes

  c. $x = 3 - \dfrac{3}{2}y$
$\dfrac{3}{2}y + x = 3$
$\dfrac{3}{2}y = -x + 3$

$y = -\dfrac{2}{3}x + 2$
Yes

  d. $6 - 2x - y = 0$
$y = 6 - 2x = -2x + 6$
No

  e. $y = 2 - \dfrac{2}{3}x = -\dfrac{2}{3}x + 2$
Yes

  f. $x + y = 1$
$y = -x + 1$
No

31. **a.** $x + y = 3$
$y = -x + 3$
$m = -1, b = 3$
L_3

  b. $2x - y = -2$
$-y = -2x - 2$
$y = 2x + 2$
$m = 2, b = 2$
L_1

  c. $x = 3y + 3$
$x - 3 = 3y$
$y = \dfrac{1}{3}x - 1$
$m = \dfrac{1}{3}, b = -1$
L_2

33. **a.**   Water boils at 212°F.
$212 = 30x + 72$
$x = 4\dfrac{2}{3}$
4 minutes, 40 seconds

  b.   Temperature at $x = 0$:
72° water was placed in the kettle.

  c.   No, because the water is never at 0°.

4

35. a. x-intercept: $\left(-33\frac{1}{3}, 0\right)$

y-intercept: $(0, 2.5)$

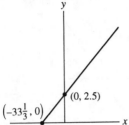

b. In 1960, 2.5 trillion cigarettes were sold.

c. $4 = .075x + 2.5$
$x = 20$
$1960 + 20 = 1980$

d. $2020 - 1960 = 60$
$y = .075(60) + 2.5$
$y = 7$
7 trillion

37. Up, because the y-intercept moves up.

39. On the x-axis, $y = 0$.

41. No, not $x = 1$, for example.

Exercises 2

1. False

3. True

5. $2x - 5 \geq 3$
$2x \geq 8$
$x \geq 4$

7. $-5x + 13 \leq -2$
$-5x \leq -15$
$x \geq 3$

9. $2x + y \leq 5$
$y \leq -2x + 5$

11. $5x - \frac{1}{3}y \leq 6$

$-\frac{1}{3}y \leq -5x + 6$

$y \geq 15x - 18$

13. $4x \geq -3$

$x \geq -\frac{3}{4}$

15. $3(2) + 5(1) \leq 12$
$6 + 5 \leq 12$
$11 \leq 12$
Yes

17. $0 \geq -2(3) + 7$
$0 \geq -6 + 7$
$0 \geq 1$
No

19. $5 \leq 3(3) - 4$
$5 \leq 9 - 4$
$5 \leq 5$
Yes

21. $7 \geq 5$
Yes

23.

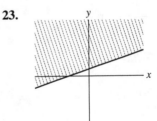

25.

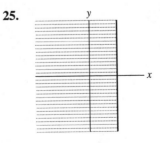

27.

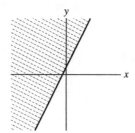

29.

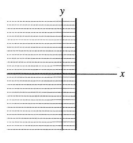

31. $x + 4y \geq 12$

$y \geq -\dfrac{1}{4}x + 3$

33. $4x - 5y + 25 \geq 0$

$y \leq \dfrac{4}{5}x + 5$

35. $\dfrac{1}{2}x - \dfrac{1}{3}y \leq 1$

$y \geq \dfrac{3}{2}x - 3$

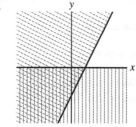

37.

39. $\begin{cases} x + 2y \geq 2 \\ 3x - y \geq 3 \end{cases}$

$\begin{cases} y \geq -\dfrac{1}{2}x + 1 \\ y \leq 3x - 3 \end{cases}$

41. $\begin{cases} x + 5y \le 10 \\ x + y \le 3 \\ x \ge 0,\ y \ge 0 \end{cases}$

$\begin{cases} y \le -\dfrac{1}{5}x + 2 \\ y \le -x + 3 \\ x \ge 0,\ y \ge 0 \end{cases}$

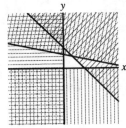

43. $\begin{cases} 6(8) + 3(7) \le 96 \\ 8 + 7 \le 18 \\ 2(8) + 6(7) \le 72 \\ 8 \ge 0,\ 7 \ge 0 \end{cases}$

$\begin{cases} 69 \le 96 \\ 15 \le 18 \\ 58 \le 72 \\ 8 \ge 0,\ 7 \ge 0 \end{cases}$

Yes

45. $\begin{cases} 6(9) + 3(10) \le 96 \\ 9 + 10 \le 18 \\ 2(9) + 6(10) \le 72 \\ 9 \ge 0,\ 10 \ge 0 \end{cases}$

$\begin{cases} 84 \le 96 \\ 19 \le 18 \qquad \mathbf{X} \\ 78 \le 72 \qquad \mathbf{X} \\ 9 \ge 0,\ 10 \ge 0 \end{cases}$

No

47. For $x = 3$, $y = 2(3) + 5 = 11$.
So $(3, 9)$ is below.

49. $7 - 4x + 5y = 0$

$y = \dfrac{4}{5}x - \dfrac{7}{5}$

For $x = 0$, $y = \dfrac{4}{5}(0) - \dfrac{7}{5} = -\dfrac{7}{5}$.

So $(0, 0)$ is above.

51. $8x - 4y - 4 = 0$
$y = 2x - 1$
$8x - 4y = 0$
$y = 2x$
$\begin{cases} y \ge 2x - 1 \\ y \le 2x \end{cases}$

53. If $x > 0$ and $y < 0$, $2x - 3y > -6$.
Quadrant IV, or (e)

55. $x + 2y = 11$

$y = -\dfrac{1}{2}x + \dfrac{11}{2}$

a.

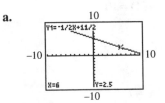

$(6, 2.5)$

b. Above, because $(6, 2.5)$ is on the line.

57. $\begin{cases} 3x + 6y \ge 24 \\ 3x + y \ge 6 \end{cases}$

$\begin{cases} y \ge -\dfrac{1}{2}x + 4 \\ y \ge -3x + 6 \end{cases}$

Exercises 3

1. $4x - 5 = -2x + 7$
   $6x = 12$
   $x = 2$
   $y = 4(2) - 5 = 3$
   $(2, 3)$

3. $x = 4y - 2$
   $x = -2y + 4$
   $4y - 2 = -2y + 4$
   $6y = 6$
   $y = 1$
   $x = 4(1) - 2 = 2$
   $(2, 1)$

5. $y = \dfrac{1}{3}(12) - 1 = 3$
   $(12, 3)$

7. $\begin{cases} 6 - 3(4) = -6 \\ 3(6) - 2(4) = 10 \end{cases}$
   $\begin{cases} -6 = -6 \\ 10 = 10 \end{cases}$
   Yes

9. $\begin{cases} y = -2x + 7 \\ y = x - 3 \end{cases}$
   $-2x + 7 = x - 3$
   $-3x = -10$
   $x = \dfrac{10}{3}$
   $y = \dfrac{10}{3} - 3 = \dfrac{1}{3}$
   $x = \dfrac{10}{3},\ y = \dfrac{1}{3}$

11. $\begin{cases} y = \dfrac{5}{2}x - \dfrac{1}{2} \\ y = -2x - 4 \end{cases}$
    $\dfrac{5}{2}x - \dfrac{1}{2} = -2x - 4$
    $\dfrac{9}{2}x = -\dfrac{7}{2}$
    $x = -\dfrac{7}{9}$

$y = -2\left(-\dfrac{7}{9}\right) - 4 = -\dfrac{22}{9}$

$x = -\dfrac{7}{9},\ y = -\dfrac{22}{9}$

13. $\begin{cases} x = 3 \\ 2x + 3y = 18 \end{cases}$
    $y = -\dfrac{2}{3}x + 6 = -\dfrac{2}{3}(3) + 6 = 4$
    $A = (3, 4)$
    $\begin{cases} x = 2 \\ 2x + 3y = 18 \end{cases}$
    $x = -\dfrac{3}{2}y + 9 = -\dfrac{3}{2}(2) + 9 = 6$
    $B = (6, 2)$

15. $A = (0, 0)$
    $\begin{cases} y = 2x \\ y = \dfrac{1}{2}x + 3 \end{cases}$
    $2x = \dfrac{1}{2}x + 3$
    $x = 2$
    $y = 2(2) = 4$
    $B = (2, 4)$
    $\begin{cases} y = \dfrac{1}{2}x + 3 \\ x = 5 \end{cases}$
    $y = \dfrac{1}{2}(5) + 3 = \dfrac{11}{2}$
    $C = \left(5,\ \dfrac{11}{2}\right)$
    $D = (5, 0)$

17. $\begin{cases} 2y - x \le 6 \\ x + 2y \ge 10 \\ \quad x \le 6 \end{cases}$

$\begin{cases} y \le \dfrac{1}{2}x + 3 \\ y \ge -\dfrac{1}{2}x + 5 \\ x \le 6 \end{cases}$

$$\begin{cases} y = \dfrac{1}{2}x + 3 \\ y = -\dfrac{1}{2}x + 5 \end{cases} \Rightarrow (2, 4)$$

$$\begin{cases} y = -\dfrac{1}{2}x + 5 \\ x = 6 \end{cases} \Rightarrow (6, 2)$$

$$\begin{cases} y = \dfrac{1}{2}x + 3 \\ x = 6 \end{cases} \Rightarrow (6, 6)$$

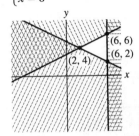

19. $\begin{cases} x + 3y \le 18 \\ 2x + y \le 16 \\ x \ge 0,\ y \ge 0 \end{cases}$

$\begin{cases} y \le -\dfrac{1}{3}x + 6 \\ y \le -2x + 16 \\ x \ge 0,\ y \ge 0 \end{cases}$

$\begin{cases} y = -\dfrac{1}{3}x + 6 \\ y = -2x + 16 \end{cases} \Rightarrow (6, 4)$

$\begin{cases} y = -\dfrac{1}{3}x + 6 \\ x = 0 \end{cases} \Rightarrow (0, 6)$

$\begin{cases} y = -2x + 16 \\ y = 0 \end{cases} \Rightarrow (8, 0)$

$$\begin{cases} x = 0 \\ y = 0 \end{cases} \Rightarrow (0, 0)$$

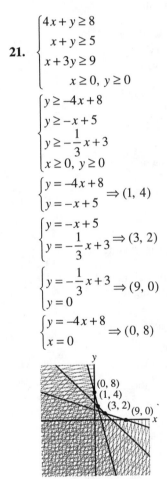

21. $\begin{cases} 4x + y \ge 8 \\ x + y \ge 5 \\ x + 3y \ge 9 \\ x \ge 0,\ y \ge 0 \end{cases}$

$\begin{cases} y \ge -4x + 8 \\ y \ge -x + 5 \\ y \ge -\dfrac{1}{3}x + 3 \\ x \ge 0,\ y \ge 0 \end{cases}$

$\begin{cases} y = -4x + 8 \\ y = -x + 5 \end{cases} \Rightarrow (1, 4)$

$\begin{cases} y = -x + 5 \\ y = -\dfrac{1}{3}x + 3 \end{cases} \Rightarrow (3, 2)$

$\begin{cases} y = -\dfrac{1}{3}x + 3 \\ y = 0 \end{cases} \Rightarrow (9, 0)$

$\begin{cases} y = -4x + 8 \\ x = 0 \end{cases} \Rightarrow (0, 8)$

23. a. $q = 10{,}000(2) - 500$
$ = 19{,}500 \text{ units}$

b. $0 = 10,000p - 500$
$p = 0.05$
5 cents

25. a. $\begin{cases} q = 10,000p - 500 \\ q = -1000p + 32,500 \end{cases}$
$10,000p - 500 = -1000p + 32,500$
$11,000p = 33,000$
$p = 3$ dollars

b. $q = 10,000(3) - 500$
$= 29,500$ units

27. Let x = hours working and
y = hours supervising.
$\begin{cases} x + y = 40 \\ 12x + 15y = 504 \end{cases}$
$\begin{cases} y = -x + 40 \\ y = -\dfrac{4}{5}x + \dfrac{168}{5} \end{cases}$
$-x + 40 = -\dfrac{4}{5}x + \dfrac{168}{5}$
$-\dfrac{1}{5}x = -\dfrac{32}{5}$
$x = 32$
$y = -32 + 40 = 8$
Working: 32; supervising: 8

29. Let x = first lift weight and
y = second lift weight.
$\begin{cases} x + y = 750 \\ 2x = y + 300 \end{cases}$
$\begin{cases} y = -x + 750 \\ y = 2x - 300 \end{cases}$
$-x + 750 = 2x - 300$
$-3x = -1050$
$x = 350$
(d)

31.

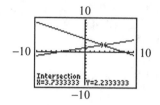

(3.73, 2.23)

33. $\begin{cases} x - 4y = -5 \\ 3x - 2y = 4.2 \end{cases}$
$\begin{cases} y = \dfrac{1}{4}x + \dfrac{5}{4} \\ y = \dfrac{3}{2}x - 2.1 \end{cases}$

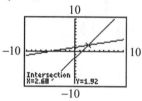

(2.68, 1.92)

35. $\begin{cases} -x + 3y \ge 3 \\ .4x + y \ge 3.2 \end{cases}$
$\begin{cases} y \ge \dfrac{1}{3}x + 1 \\ y \ge -.4x + 3.2 \end{cases}$

a.

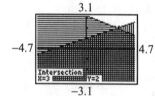

b. (3, 2)

c.

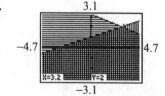

d. $\begin{cases} -(32)+3(2) \geq 3 \\ .4(3.2)+2 \geq 3.2 \end{cases}$

$\begin{cases} 2.8 \geq 3 \\ 3.28 \geq 3.2 \end{cases}$

No

Exercises 4

1. $m = \dfrac{2}{3}$

3. $y - 3 = 5(x + 4)$
$y = 5x + 23$
$m = 5$

5.

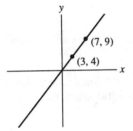

$m = \dfrac{9-4}{7-3} = \dfrac{5}{4}$

7.

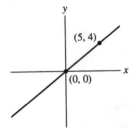

$m = \dfrac{4-0}{5-0} = \dfrac{4}{5}$

9. The slope of a vertical line is undefined.

11.

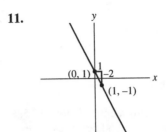

13.

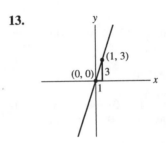

15. $m = \dfrac{-2}{1} = -2$
$y - 3 = -2(x - 2)$
$y = -2x + 7$

17. $m = \dfrac{0-2}{2-1} = -2$
$y - 0 = -2(x - 2)$
$y = -2x + 4$

19. $m = -\dfrac{1}{-4} = \dfrac{1}{4}$

$y - 2 = \dfrac{1}{4}(x - 2)$

$y = \dfrac{1}{4}x + \dfrac{3}{2}$

21. $m = -1$
$y - 0 = -1(x - 0)$
$y = -x$

23. $m = 0$
$y - 3 = 0(x - 2)$
$y = 3$

25. $y - 6 = \dfrac{3}{5}(x - 5)$

$y = \dfrac{3}{5}x + 3$

y-intercept: (0, 3)

27. Each unit sold yields a commission of $5. In addition, she receives $60 per week base pay.

29. a. p-intercept: (0, 1200); at $1200 no one will buy the item.

b. $0 = -3q + 1200$
$q = 400$ units
q-intercept: (400, 0); even if the item is given away, only 400 will be taken.

c. -3; to sell an additional item, the price must be reduced by $3.

d. $p = -3(350) + 1200 = \$150$

e. $300 = -3q + 1200$
$q = 300$ items

f.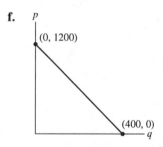

31. a. Let x = quantity and y = cost.
$m = \dfrac{9500 - 6800}{50 - 20} = 90$
$y - 6800 = 90(x - 20)$
$y = 90x + 5000$

b. $5000

c. $90

d.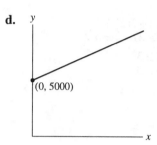

33. a. $300(100) = \$30,000$

b. $6000 = 100x$
$x = 60$ coats

c. $y = 100(0) = 0$
(0, 0); if no coats are sold, there is no revenue.

d. 100; each additional coat yields an additional $100 in revenue.

35.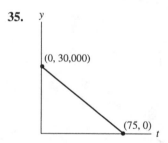

37. $y = 30,000 - 400(45)$
$= 12,000$ gallons

39. $0 = 30,000 - 400t$
$t = 75$
(75, 0)
The tank is empty after 75 days.

41. $y = 2.3 - .15(15) = \$.05$ million
$50,000

12

43. $0 = 2.3 - .15t$

$t = 15\dfrac{1}{3}$

$\left(15\dfrac{1}{3}, 0\right)$; the cash reserves will be

depleted after $15\dfrac{1}{3}$ days.

45. $.8 = 2.3 - .15t$
$t = 10$
After 10 days, on July 11

47. $y = \dfrac{1}{200}(0) + 3 = 3$

49. $10 = \dfrac{1}{200}x + 3$
$x = 1400$ hours

51. $y = .10(1000) + 160$
$= \$260$

53. $m = 3, b = -1$
$y = 3x - 1$

55. $m = 1$
$y - 2 = 1(x - 1)$
$y = x + 1$

57. $m = -7$
$y - 0 = -7(x - 5)$
$y = -7x + 35$

59. $m = 0$
$y - 4 = 0(x - 7)$
$y = 4$

61. $m = \dfrac{2 - 1}{4 - 2} = \dfrac{1}{2}$

$y - 1 = \dfrac{1}{2}(x - 2)$

$y = \dfrac{1}{2}x$

63. $m = \dfrac{-2 - 0}{1 - 0} = -2$
$y = -2x$

65. Changes in x-coordinate: $1, -1, -2$
Changes in y-coordinate are m times that:
$2, -2, -4$
y-coordinates:
$3 + 2 = 5, 3 - 2 = 1, 3 - 4 = -1$
$5; 1; -1$

67. Changes in x-coordinate:
$1, 2, -1$
Changes in y-coordinate are m times that:
$-\dfrac{1}{4}, -\dfrac{1}{2}, \dfrac{1}{4}$
y-coordinates:
$-1 - \dfrac{1}{4} = -\dfrac{5}{4}, \; -1 - \dfrac{1}{2} = -\dfrac{3}{2}, \; -1 + \dfrac{1}{4} = -\dfrac{3}{4}$
$-\dfrac{5}{4}; -\dfrac{3}{2}; -\dfrac{3}{4}$

69. a. $x + y = 1$
$y = -x + 1$
(C)

b. $x - y = 1$
$y = x - 1$
(B)

c. $x + y = -1$
$y = -x - 1$
(D)

d. $x - y = -1$
$y = x + 1$
(A)

71. $m = \dfrac{212 - 32}{100 - 0} = \dfrac{9}{5}$

$F - 32 = \dfrac{9}{5}(C - 0)$

$F = \dfrac{9}{5}C + 32$

73. $y = $ millions of farms
$m = \dfrac{2 - 6}{60} = -\dfrac{1}{15}$

$y = -\dfrac{1}{15}x + 6$

$y = -\dfrac{1}{15}(45) + 6 = 3$ million

75. Counterclockwise

77. $y \geq 4x + 3$

79. $m_1 = \dfrac{3-4}{2-0} = -\dfrac{1}{2}$

$y = -\dfrac{1}{2}x + 4$

$m_2 = \dfrac{1-3}{4-2} = -1$

$y - 1 = -(x-4)$

$y = -x + 5$

$m_3 = \dfrac{1-0}{4-3} = 1$

$y = x - 3$

$\begin{cases} y \leq -\dfrac{1}{2}x + 4 \\ y \leq -x + 5 \\ y \geq x - 3 \\ x \geq 0, \ y \geq 0 \end{cases}$

81. Set two slopes equal:

$\dfrac{7-5}{2-1} = \dfrac{k-7}{3-2}$

$2 = k - 7$

$k = 9$

83. Make slopes negative inverses of each other:

$\dfrac{-3.1-1}{2-a} = -\dfrac{1}{\frac{2.4-0}{3.8-(-1)}}$

$\dfrac{-4.1}{2-a} = -2$

$4.1 = 4 - 2a$

$a = -.05$

85. $l_1: y = m_1 x$

$l_2: y = m_2 x$

So the vertical segment lies on $x = 1$.

Then

$1^2 + m_1^2 = a^2$

$1^2 + m_2^2 = b^2$

Add equations and rearrange:

$a^2 + b^2 - (m_1^2 + m_2^2) = 2$

l_1 and l_2 are perpendicular if and only if

$a^2 + b^2 = (m_1 - m_2)^2 = m_1^2 + m_2^2 - 2m_1 m_2$

or

$a^2 + b^2 - (m_1^2 + m_2^2) = -2m_1 m_2$

Substitute:

$2 = -2m_1 m_2$

$m_2 = -\dfrac{1}{m_1}$

87. $m = \dfrac{75,000 - 15,000}{20,000 - 0} = 3$

$y = 3x + 15,000$

$= 3(12,000) + 15,000 = \$51,000$

(c)

89. $n = 300 - 20(10) = 100$

Revenue: $pn = 10 \cdot 100 = \$1000$

(c)

91.

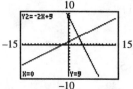

No, do not appear perpendicular

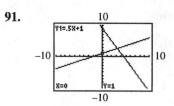

Do appear perpendicular

93.

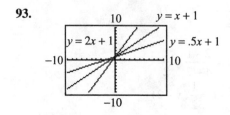

Use the slopes.

Steepest: $y = 2x + 1$

Middle: $y = x + 1$

Flattest: $y = .5x + 1$

95.

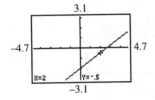

$$.75 = m = \frac{x}{2}$$
$$x = 1.5$$
Up 1.5 units

Exercises 5

1.

Data Point	Point on Line	Vertical Distance
(1, 3)	(1, 4)	1
(2, 6)	(2, 7)	1
(3, 11)	(3, 10)	1
(4, 12)	(4, 13)	1

$$1^2 + 1^2 + 1^2 + 1^2 = 4$$

3. $E_1^2 = [1.1(1) + 3 - 3]^2 = 1.21$
$E_2^2 = [1.1(2) + 3 - 6]^2 = 0.64$
$E_3^2 = [1.1(3) + 3 - 8]^2 = 2.89$
$E_4^2 = [1.1(4) + 3 - 6]^2 = 1.96$
$E = 1.21 + 0.64 + 2.89 + 1.96 = 6.70$

5.

x	y	xy	x^2
1	7	7	1
2	6	12	4
3	4	12	9
4	3	12	16
$\sum x = 10$	$\sum y = 20$	$\sum xy = 43$	$\sum x^2 = 30$

$$m = \frac{4 \cdot 43 - 10 \cdot 20}{4 \cdot 30 - 10^2} = -1.4$$
$$b = \frac{20 - (-1.4)(10)}{4} = 8.5$$

7. $\sum x = 6, \sum y = 18, \sum xy = 45,$
$\sum x^2 = 14$

$m = \dfrac{3 \cdot 45 - 6 \cdot 18}{3 \cdot 14 - 6^2} = 4.5$

$b = \dfrac{18 - (4.5)(6)}{3} = -3$

$y = 4.5x - 3$

9. $\sum x = 10, \sum y = 26, \sum xy = 55,$
$\sum x^2 = 30$

$m = \dfrac{4 \cdot 55 - 10 \cdot 26}{4 \cdot 30 - 10^2} = -2$

$b = \dfrac{26 - (-2)(10)}{4} = 11.5$

$y = -2x + 11.5$

11. a.
```
LinReg
y=ax+b
a=.3383317713
b=21.62136832
```

$y = .338x + 21.6$

b. $.338(1100) + 21.6 = 393.4$
About 393 deaths per million males

13. a. Let $x =$ years since 1970 and $y =$ percent.
```
LinReg
y=ax+b
a=.4108571429
b=11.78095238
```

$y = .411x + 11.8$

b. $.497(23) + 11.2 = 21.253$
About 21.3 percent

c. $27.1 = .411x + 11.8$
$x \approx 37.23$
The year 2007

15. a.
```
LinReg
y=ax+b
a=.6077890746
b=35.40003446
```

$y = .608x + 35.4$

b. $.608(74) + 35.4 = 80.392$
About 80.4 years

c. $77.7 = .608x + 35.4$
$x \approx 69.572$
About 69.6 years

17. Yes, because E will remain the same and can only be increased by moving the line.

Chapter 1 Supplementary Exercises

1. $x = 0$

2.

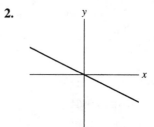

3. $\begin{cases} x - 5y = 6 \\ \quad 3x = 6 \end{cases}$
$\begin{cases} x = 5y + 6 \\ x = 2 \end{cases}$
$5y + 6 = 2$
$y = -\dfrac{4}{5}$
$\left(2, -\dfrac{4}{5}\right)$

4. $3x - 4y = 8$
$y = \dfrac{3}{4}x - 2$
$m = \dfrac{3}{4}$

5. $m = \dfrac{0-5}{10-0} = -\dfrac{1}{2}, \; b = 5$

$y = -\dfrac{1}{2}x + 5$

6. $x - 3y \geq 12$

$y \leq \dfrac{1}{3}x - 4$

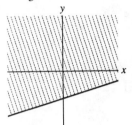

7. $3(1) + 4(2) \geq 11$
$3 + 8 \geq 11$
$11 \geq 11$
Yes

8. $\begin{cases} 2x - y = 1 \\ x + 2y = 13 \end{cases}$

$\begin{cases} y = 2x - 1 \\ y = -\dfrac{1}{2}x + \dfrac{13}{2} \end{cases}$

$2x - 1 = -\dfrac{1}{2}x + \dfrac{13}{2}$

$\dfrac{5}{2}x = \dfrac{15}{2}$

$x = 3$
$y = 2(3) - 1 = 5$
$(3, 5)$

9. $2x - 10y = 7$

$y = \dfrac{1}{5}x - \dfrac{7}{10}$

$m = \dfrac{1}{5}$

$y - 16 = \dfrac{1}{5}(x - 15)$

$y = \dfrac{1}{5}x + 13$

10. $y = 3(1) + 7 = 10$

11. $(5, 0)$

12.

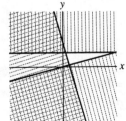

13. $\begin{cases} 3x - 2y = 1 \\ 2x + y = 24 \end{cases}$

$\begin{cases} y = \dfrac{3}{2}x - \dfrac{1}{2} \\ y = -2x + 24 \end{cases}$

$\dfrac{3}{2}x - \dfrac{1}{2} = -2x + 24$

$\dfrac{7}{2}x = \dfrac{49}{2}$

$x = 7$
$y = -2(7) + 24 = 10$
$(7, 10)$

14. $\begin{cases} 2y + 7x \geq 30 \\ 4y - x \geq 0 \\ \quad y \leq 8 \end{cases}$

$\begin{cases} y \geq -\dfrac{7}{2}x + 15 \\ y \geq \dfrac{1}{4}x \\ y \leq 8 \end{cases}$

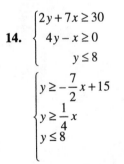

15. $y - 9 = \frac{1}{2}(x - 4)$

$y = \frac{1}{2}x + 7$

$b = 7$

$(0, 7)$

16. The rate is $35 per hour plus a flat fee of $20.

17. $m_1 = \frac{0-2}{2-1} = -2$

$m_2 = \frac{1-0}{3-2} = 1$

$m_1 \neq m_2$

No

18. $m = \frac{-2-0}{0-3} = \frac{2}{3}$, $b = -2$

$y = \frac{2}{3}x - 2$

19. $\begin{cases} x + 5y = 16 \\ x = -3y \end{cases}$

$\begin{cases} x = -5y + 16 \\ x = -3y \end{cases}$

$-5y + 16 = -3y$

$16 = 2y$

$y = 8$

(e)

20. $y \leq \frac{2}{3}x + \frac{3}{2}$

21. $m = \frac{8.6 - (-1)}{6 - 2} = 2.4$

$y - (-1) \geq 2.4(x - 2)$

$y \geq 2.4x - 5.8$

22. $\begin{cases} 1.2x + 2.4y = .6 \\ 4.8y - 1.6x = 2.4 \end{cases}$

$\begin{cases} y = -.5x + .25 \\ y = \frac{1}{3} + .5 \end{cases}$

$-.5x + .25 = \frac{1}{3}x + .5$

$-\frac{5}{6}x = .25$

$x = -.3$

$y = \frac{1}{3}(-.3) + .5 = .4$

23. $\begin{cases} y = -x + 1 \\ y = 2x + 3 \end{cases}$

$-x + 1 = 2x + 3$

$-3x = 2$

$x = -\frac{2}{3}$

$y = -\left(-\frac{2}{3}\right) + 1 = \frac{5}{3}$

$\left(-\frac{2}{3}, \frac{5}{3}\right)$

$m = \frac{\frac{5}{3} - 1}{-\frac{2}{3} - 1} = -\frac{2}{5}$

$y - 1 = -\frac{2}{5}(x - 1)$

$y = -\frac{2}{5}x + \frac{7}{5}$

24. $2x + 3(x - 2) \geq 0$

$5x \geq 6$

$x \geq \frac{6}{5}$

25. $x + \frac{1}{2}y = 4$

$y = -2x + 8$

$m = -2$

y-intercept: $(0, 8)$

$0 = -2x + 8$

$x = 4$

x-intercept: $(4, 0)$

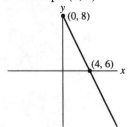

26. $\begin{cases} 5x + 2y = 0 \\ \quad x + y = 1 \end{cases}$

$\begin{cases} y = -\dfrac{5}{2}x \\ y = -x + 1 \end{cases}$

$-\dfrac{5}{2}x = -x + 1$

$-\dfrac{3}{2}x = 1$

$x = -\dfrac{2}{3}$

$y = -\left(-\dfrac{2}{3}\right) + 1 = \dfrac{5}{3}$

Substitute $x = -\dfrac{2}{3}$ and $y = \dfrac{5}{3}$ in

$2x - 3y = 1$.

$2\left(-\dfrac{2}{3}\right) - 3\left(\dfrac{5}{3}\right) = 1$

$-\dfrac{19}{3} = 1$ **X**

No

27. $\begin{cases} 2x - 3y = 1 \\ 3x + 2y = 4 \end{cases}$

$\begin{cases} y = \dfrac{2}{3}x - \dfrac{1}{3} \\ y = -\dfrac{3}{2}x + 2 \end{cases}$

$m_1 = -\dfrac{1}{m_2}$

28. a. $x + y \geq 1$
$y \geq -x + 1$
(C)

    b. $x + y \leq 1$
$y \leq -x + 1$
(A)

    c. $x - y \leq 1$
$y \geq x - 1$
(B)

    d. $y - x \leq -1$
$y \leq x - 1$
(D)

29. a. $4x + y = 17$
$y = -4x + 17$
L_3

    b. $y = x + 2$
L_1

    c. $2x + 3y = 11$
$y = -\dfrac{2}{3}x + \dfrac{11}{3}$
L_2

30. $m_1 = \dfrac{\frac{3}{2} - 5}{4 - 0} = -\dfrac{7}{8}, \; b_1 = 5$

$y = -\dfrac{7}{8}x + 5$

$m_2 = -\dfrac{1}{m_1} = \dfrac{8}{7}$

$y - \dfrac{3}{2} = \dfrac{8}{7}(x - 4)$

$y = \dfrac{8}{7}x - \dfrac{43}{14}$

$\begin{cases} y \leq -\dfrac{7}{8}x + 5 \\ y \geq \dfrac{8}{7}x - \dfrac{43}{14} \\ x \geq 0, \; y \geq 0 \end{cases}$

$0 = \dfrac{8}{7}x - \dfrac{43}{14}$

$x = \dfrac{43}{16}$

$\left(\dfrac{43}{16}, \, 0\right)$

31. Supply curve: $q = 150p - 100$
(positive slope, negative y-intercept)
Demand curve: $q = -75p + 500$
(negative slope, positive y-intercept)
$150p - 100 = -75p + 500$
$225p = 600$
$p = \dfrac{8}{3}$

$$q = 150\left(\frac{8}{3}\right) - 100 = 300$$

$$\left(\frac{8}{3}, 300\right)$$

32. $\begin{cases} x \geq 0 \\ y \geq 0 \end{cases}$

$(0, 0)$

$\begin{cases} y \geq 0 \\ 5x + y \leq 50 \end{cases}$

$\begin{cases} y \geq 0 \\ y \leq -5x + 50 \end{cases}$

$0 = -5x + 50$

$x = 10$

$(10, 0)$

$\begin{cases} 5x + y \leq 50 \\ 2x + 3y \leq 33 \end{cases}$

$\begin{cases} y \leq -5x + 50 \\ y \leq -\dfrac{2}{3}x + 11 \end{cases}$

$-5x + 50 = -\dfrac{2}{3}x + 11$

$-\dfrac{13}{3}x = -39$

$x = 9$

$y = -5(9) + 50 = 5$

$(9, 5)$

$\begin{cases} 2x + 3y \leq 33 \\ x - 2y \geq -8 \end{cases}$

$\begin{cases} y \leq -\dfrac{2}{3}x + 11 \\ y \leq \dfrac{1}{2}x + 4 \end{cases}$

$-\dfrac{2}{3}x + 11 = \dfrac{1}{2}x + 4$

$-\dfrac{7}{6}x = -7$

$x = 6$

$x = \dfrac{1}{2}(6) + 4 = 7$

$(6, 7)$

$\begin{cases} x - 2y \geq -8 \\ x \geq 0 \end{cases}$

$\begin{cases} x \geq 2y - 8 \\ x \geq 0 \end{cases}$

$2y - 8 = 0$

$y = 4$

$(0, 4)$

33. $\begin{cases} d_1 = 2m + b - 4 \\ d_2 = 8 - (5m + b) \\ d_3 = 7m + b - 9 \\ d_1 = d_2 = d_3 \end{cases}$

$2m + b - 4 = 7m + b - 9$

$-5m = -5$

$m = 1$

$8 - (5 \cdot 1 + b) = 7 \cdot 1 + b - 9$

$3 - b = -2 + b$

$b = \dfrac{5}{2}$

$y = x + \dfrac{5}{2}$

34. a. $m = 10$

$y - 4000 = 10(x - 1000)$

$y = 10x - 6000$

b. $0 = 10x - 6000$

$x = 600$

x-intercept: $(600, 0)$

y-intercept: $(0, -6000)$

c.

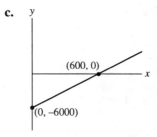

35. a. A: $y = .1x + 50$

B: $y = .2x + 40$

b. A: $.1(80) + 50 = 58$

B: $.2(80) + 40 = 56$

Company B

c. A: $.1(160) + 50 = 66$

B: $.2(160) + 40 = 72$

Company A

d. $.1x + 50 = .2x + 40$
$-.1x = -10$
$x = 100$ miles

36. a. $m = \dfrac{-4000}{8} = -500, \ b = 5000$
$y = -500x + 5000$

b. $y = -500(4) + 5000 = \$3000$

c. $2000 = -500x + 5000$
$500x = 3000$
$x = 6$
1992

37. $x \le 3y + 2$
$y \ge \dfrac{1}{3}x - \dfrac{2}{3}$

38. $.03x + 200 = .05x + 100$
$-.02x = -100$
$x = \$5000$

39. $m_1 = \dfrac{5 - 0}{0 - (-4)} = \dfrac{5}{4}, \ b_1 = 5$

$y = \dfrac{5}{4}x + 5$

$m_2 = \dfrac{0 - 2}{5 - 0} = -\dfrac{2}{5}, \ b_2 = 2$

$y = -\dfrac{2}{5}x + 2$

$m_3 = \dfrac{0 - (-3)}{5 - 0} = \dfrac{3}{5}, \ b_3 = -3$

$y = \dfrac{3}{5}x - 3$

$m_4 = \dfrac{-5 - 0}{0 - (-2)} = -\dfrac{5}{2}, \ b_4 = -5$

$y = -\dfrac{5}{2}x - 5$

$\begin{cases} y \le \dfrac{5}{4}x + 5 \\ y \le -\dfrac{2}{5}x + 2 \\ y \ge \dfrac{3}{5}x - 3 \\ y \ge -\dfrac{5}{2} - 5 \end{cases}$

40. $m_1 = \dfrac{2 - 0}{0 - 4} = -\dfrac{1}{2}, \ b_1 = 2$

$y = -\dfrac{1}{2}x + 2$
The other lines are $x = -2$, $x = 4$, and $y = -3$.

$\begin{cases} y \le -\dfrac{1}{2}x + 2 \\ x \ge -2 \\ x \le 4 \\ y \ge -3 \end{cases}$

41. a. Let x = years since 1980 and
y = number of degrees (thousands).

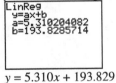

```
LinReg
y=ax+b
a=5.310204082
b=193.8285714
```

$y = 5.310x + 193.829$

b. $5.310(15) + 193.829 = 273.479$
About 273.5 thousand

c. $300 = 5.310x + 193.829$
$x \approx 19.99$
The year 2000

42. a.
```
LinReg
y=ax+b
a=.1517702501
b=-3.063197325
```

$y = .152x - 3.063$

b. $.152(160) - 3.063 = 21.257$
About 21 deaths per 100,000

c. $22 = .152x - 3.063$
$x \approx 164.888$
About 165 grams

Chapter 2

1. $\begin{cases} \dfrac{1}{2}x - 3y = 2 \\ 5x + 4y = 1 \end{cases}$

$\xrightarrow{\;2[1]\;} \begin{cases} x - 6y = 4 \\ 5x + 4y = 1 \end{cases}$

3. $\quad 5\ \text{(first)}\quad 5x + 10y = 15$

$\quad +\text{(second)}\quad \dfrac{-5x + 4y = 1}{14y = 16}$

$\begin{cases} x + 2y = 3 \\ -5x + 4y = 1 \end{cases}$

$\xrightarrow{\;[2]+5[1]\;} \begin{cases} x + 2y = 3 \\ 14y = 16 \end{cases}$

5. $\quad -4\text{(first)}\quad -4x + 8y - 4z = 0$

$\quad +\text{(third)}\quad \dfrac{4x + y + 3z = 5}{9y - z = 5}$

$\begin{cases} x - 2y + z = 0 \\ y - 2z = 4 \\ 4x + y + 3z = 5 \end{cases}$

$\xrightarrow{\;[3]+(-4)[1]\;} \begin{cases} x - 2y + z = 0 \\ y - 2z = 4 \\ 9y - z = 5 \end{cases}$

7. $\begin{bmatrix} 1 & -\frac{1}{2} & 3 \\ 0 & 1 & 4 \end{bmatrix} \xrightarrow{\;[1]+\frac{1}{2}[2]\;} \begin{bmatrix} 1 & 0 & 5 \\ 0 & 1 & 4 \end{bmatrix}$

9. Use [2] + 2[1] to change the –2 to a 0.

11. Use [1] to (–2)[2] to change the 2 to a 0.

13. Interchange rows 1 and 2 or rows 1 and 3 to make the first entry in row 1 non-zero.

15. Use [1] + (–3)[3] to change the 3 to a 0.

17. $\begin{bmatrix} 3 & 9 & 6 \\ 2 & 8 & 6 \end{bmatrix} \xrightarrow{\;\frac{1}{3}[1]\;} \begin{bmatrix} 1 & 3 & 2 \\ 2 & 8 & 6 \end{bmatrix}$

$\xrightarrow{\;[2]+(-2)[1]\;} \begin{bmatrix} 1 & 3 & 2 \\ 0 & 2 & 2 \end{bmatrix}$

$\xrightarrow{\;\frac{1}{2}[2]\;} \begin{bmatrix} 1 & 3 & 2 \\ 0 & 1 & 1 \end{bmatrix}$

$\xrightarrow{\;[1]+(-3)[2]\;} \begin{bmatrix} 1 & 0 & -1 \\ 0 & 1 & 1 \end{bmatrix}$

$x = -1,\ y = 1$

19. $\begin{bmatrix} 1 & -3 & 4 & 1 \\ 4 & -10 & 10 & 4 \\ -3 & 9 & -5 & -6 \end{bmatrix}$

$\xrightarrow{\;[2]+(-4)[1]\;} \begin{bmatrix} 1 & -3 & 4 & 1 \\ 0 & 2 & -6 & 0 \\ -3 & 9 & -5 & -6 \end{bmatrix}$

$\xrightarrow{\;[3]+3[1]\;} \begin{bmatrix} 1 & -3 & 4 & 1 \\ 0 & 2 & -6 & 0 \\ 0 & 0 & 7 & -3 \end{bmatrix}$

$\xrightarrow{\;\frac{1}{2}[2]\;} \begin{bmatrix} 1 & -3 & 4 & 1 \\ 0 & 1 & -3 & 0 \\ 0 & 0 & 7 & -3 \end{bmatrix}$

$\xrightarrow{\;[1]+3[2]\;} \begin{bmatrix} 1 & 0 & -5 & 1 \\ 0 & 1 & -3 & 0 \\ 0 & 0 & 7 & -3 \end{bmatrix}$

$\xrightarrow{\;\frac{1}{7}[3]\;} \begin{bmatrix} 1 & 0 & -5 & 1 \\ 0 & 1 & -3 & 0 \\ 0 & 0 & 1 & -\frac{3}{7} \end{bmatrix}$

$\xrightarrow{\;[1]+5[3]\;} \begin{bmatrix} 1 & 0 & 0 & -\frac{8}{7} \\ 0 & 1 & -3 & 0 \\ 0 & 0 & 1 & -\frac{3}{7} \end{bmatrix}$

$$\xrightarrow{[2]+3[3]} \begin{bmatrix} 1 & 0 & 0 & \bigm| & -\frac{8}{7} \\ 0 & 1 & 0 & \bigm| & -\frac{9}{7} \\ 0 & 0 & 1 & \bigm| & -\frac{3}{7} \end{bmatrix}$$

$$x = -\frac{8}{7},\; y = -\frac{9}{7},\; z = -\frac{3}{7}$$

21. $\begin{bmatrix} 2 & -2 & \bigm| & -4 \\ 3 & 4 & \bigm| & 1 \end{bmatrix} \xrightarrow{\frac{1}{2}[1]} \begin{bmatrix} 1 & -1 & \bigm| & -2 \\ 3 & 4 & \bigm| & 1 \end{bmatrix}$

$$\xrightarrow{[2]+(-3)[1]} \begin{bmatrix} 1 & -1 & \bigm| & -2 \\ 0 & 7 & \bigm| & 7 \end{bmatrix}$$

$$\xrightarrow{\frac{1}{7}[2]} \begin{bmatrix} 1 & -1 & \bigm| & -2 \\ 0 & 1 & \bigm| & 1 \end{bmatrix}$$

$$\xrightarrow{[1]+1[2]} \begin{bmatrix} 1 & 0 & \bigm| & -1 \\ 0 & 1 & \bigm| & 1 \end{bmatrix}$$

$$x = -1,\; y = 1$$

23. $\begin{bmatrix} 4 & -4 & 4 & \bigm| & -8 \\ 1 & -2 & -2 & \bigm| & -1 \\ 2 & 1 & 3 & \bigm| & 1 \end{bmatrix}$

$$\xrightarrow{\frac{1}{4}[1]} \begin{bmatrix} 1 & -1 & 1 & \bigm| & -2 \\ 1 & -2 & -2 & \bigm| & -1 \\ 2 & 1 & 3 & \bigm| & 1 \end{bmatrix}$$

$$\xrightarrow{[2]+(-1)[1]} \begin{bmatrix} 1 & -1 & 1 & \bigm| & -2 \\ 0 & -1 & -3 & \bigm| & 1 \\ 2 & 1 & 3 & \bigm| & 1 \end{bmatrix}$$

$$\xrightarrow{[3]+(-2)[1]} \begin{bmatrix} 1 & -1 & 1 & \bigm| & -2 \\ 0 & -1 & -3 & \bigm| & 1 \\ 0 & 3 & 1 & \bigm| & 5 \end{bmatrix}$$

$$\xrightarrow{(-1)[2]} \begin{bmatrix} 1 & -1 & 1 & \bigm| & -2 \\ 0 & 1 & 3 & \bigm| & -1 \\ 0 & 3 & 1 & \bigm| & 5 \end{bmatrix}$$

$$\xrightarrow{[1]+[2]} \begin{bmatrix} 1 & 0 & 4 & \bigm| & -3 \\ 0 & 1 & 3 & \bigm| & -1 \\ 0 & 3 & 1 & \bigm| & 5 \end{bmatrix}$$

$$\xrightarrow{[3]+(-3)[2]} \begin{bmatrix} 1 & 0 & 4 & \bigm| & -3 \\ 0 & 1 & 3 & \bigm| & -1 \\ 0 & 0 & -8 & \bigm| & 8 \end{bmatrix}$$

$$\xrightarrow{\left(-\frac{1}{8}\right)[3]} \begin{bmatrix} 1 & 0 & 4 & \bigm| & -3 \\ 0 & 1 & 3 & \bigm| & -1 \\ 0 & 0 & 1 & \bigm| & -1 \end{bmatrix}$$

$$\xrightarrow{[1]+(-4)[3]} \begin{bmatrix} 1 & 0 & 0 & \bigm| & 1 \\ 0 & 1 & 3 & \bigm| & -1 \\ 0 & 0 & 1 & \bigm| & -1 \end{bmatrix}$$

$$\xrightarrow{[2]+(-3)[3]} \begin{bmatrix} 1 & 0 & 0 & \bigm| & 1 \\ 0 & 1 & 0 & \bigm| & 2 \\ 0 & 0 & 1 & \bigm| & -1 \end{bmatrix}$$

$$x = 1,\; y = 2,\; z = -1$$

25. $\begin{bmatrix} .2 & .3 & \bigm| & 4 \\ .6 & 1.1 & \bigm| & 15 \end{bmatrix}$

$$\xrightarrow{5[1]} \begin{bmatrix} 1 & 1.5 & \bigm| & 20 \\ .6 & 1.1 & \bigm| & 15 \end{bmatrix}$$

$$\xrightarrow{[2]+(-.6)[1]} \begin{bmatrix} 1 & 1.5 & \bigm| & 20 \\ 0 & .2 & \bigm| & 3 \end{bmatrix}$$

$$\xrightarrow{5[2]} \begin{bmatrix} 1 & 1.5 & \bigm| & 20 \\ 0 & 1 & \bigm| & 15 \end{bmatrix}$$

$$\xrightarrow{[1]+(-1.5)[2]} \begin{bmatrix} 1 & 0 & \bigm| & -2.5 \\ 0 & 1 & \bigm| & 15 \end{bmatrix}$$

$$x = -2.5,\; y = 15$$

27. $\begin{bmatrix} 1 & 1 & 4 & \bigm| & 3 \\ 4 & 1 & -2 & \bigm| & -6 \\ -3 & 0 & 2 & \bigm| & 1 \end{bmatrix}$

$$\xrightarrow{[2]+(-4)[1]} \begin{bmatrix} 1 & 1 & 4 & \bigm| & 3 \\ 0 & -3 & -18 & \bigm| & -18 \\ -3 & 0 & 2 & \bigm| & 1 \end{bmatrix}$$

$$\xrightarrow{[3]+3[1]} \begin{bmatrix} 1 & 1 & 4 & \bigm| & 3 \\ 0 & -3 & -18 & \bigm| & -18 \\ 0 & 3 & 14 & \bigm| & 10 \end{bmatrix}$$

$$\xrightarrow{\left(-\frac{1}{3}\right)[2]} \begin{bmatrix} 1 & 1 & 4 & \bigm| & 3 \\ 0 & 1 & 6 & \bigm| & 6 \\ 0 & 3 & 14 & \bigm| & 10 \end{bmatrix}$$

$$\xrightarrow{[1]+(-1)[2]} \begin{bmatrix} 1 & 0 & -2 & | & -3 \\ 0 & 1 & 6 & | & 6 \\ 0 & 3 & 14 & | & 10 \end{bmatrix}$$

$$\xrightarrow{[3]+(-3)[2]} \begin{bmatrix} 1 & 0 & -2 & | & -3 \\ 0 & 1 & 6 & | & 6 \\ 0 & 0 & -4 & | & -8 \end{bmatrix}$$

$$\xrightarrow{\left(-\frac{1}{4}\right)[3]} \begin{bmatrix} 1 & 0 & -2 & | & -3 \\ 0 & 1 & 6 & | & 6 \\ 0 & 0 & 1 & | & 2 \end{bmatrix}$$

$$\xrightarrow{[1]+2[3]} \begin{bmatrix} 1 & 0 & 0 & | & 1 \\ 0 & 1 & 6 & | & 6 \\ 0 & 0 & 1 & | & 2 \end{bmatrix}$$

$$\xrightarrow{[2]+(-6)[3]} \begin{bmatrix} 1 & 0 & 0 & | & 1 \\ 0 & 1 & 0 & | & -6 \\ 0 & 0 & 1 & | & 2 \end{bmatrix}$$

$x = 1,\ y = -6,\ z = 2$

29. $\begin{bmatrix} -1 & 1 & 0 & | & -1 \\ 1 & 0 & 1 & | & 4 \\ 6 & -3 & 2 & | & 10 \end{bmatrix}$

$$\xrightarrow{(-1)[1]} \begin{bmatrix} 1 & -1 & 0 & | & 1 \\ 1 & 0 & 1 & | & 4 \\ 6 & -3 & 2 & | & 10 \end{bmatrix}$$

$$\xrightarrow{[2]+(-1)[1]} \begin{bmatrix} 1 & -1 & 0 & | & 1 \\ 0 & 1 & 1 & | & 3 \\ 6 & -3 & 2 & | & 10 \end{bmatrix}$$

$$\xrightarrow{[3]+(-6)[1]} \begin{bmatrix} 1 & -1 & 0 & | & 1 \\ 0 & 1 & 1 & | & 3 \\ 0 & 3 & 2 & | & 4 \end{bmatrix}$$

$$\xrightarrow{[1]+1[2]} \begin{bmatrix} 1 & 0 & 1 & | & 4 \\ 0 & 1 & 1 & | & 3 \\ 0 & 3 & 2 & | & 4 \end{bmatrix}$$

$$\xrightarrow{[3]+(-3)[2]} \begin{bmatrix} 1 & 0 & 1 & | & 4 \\ 0 & 1 & 1 & | & 3 \\ 0 & 0 & -1 & | & -5 \end{bmatrix}$$

$$\xrightarrow{(-1)[3]} \begin{bmatrix} 1 & 0 & 1 & | & 4 \\ 0 & 1 & 1 & | & 3 \\ 0 & 0 & 1 & | & 5 \end{bmatrix}$$

$$\xrightarrow{[1]+(-1)[3]} \begin{bmatrix} 1 & 0 & 0 & | & -1 \\ 0 & 1 & 1 & | & 3 \\ 0 & 0 & 1 & | & 5 \end{bmatrix}$$

$$\xrightarrow{[2]+(-1)[3]} \begin{bmatrix} 1 & 0 & 0 & | & -1 \\ 0 & 1 & 0 & | & -2 \\ 0 & 0 & 1 & | & 5 \end{bmatrix}$$

$x = -1,\ y = -2,\ z = 5$

31. $\begin{cases} x + y = 300 \\ .1x + .15y = 38 \end{cases}$

$\begin{bmatrix} 1 & 1 & | & 300 \\ .1 & .15 & | & 38 \end{bmatrix}$

$$\xrightarrow{[2]+\left(-\frac{1}{10}\right)[1]} \begin{bmatrix} 1 & 1 & | & 300 \\ 0 & .05 & | & 8 \end{bmatrix}$$

$$\xrightarrow{20[2]} \begin{bmatrix} 1 & 1 & | & 300 \\ 0 & 1 & | & 160 \end{bmatrix}$$

$$\xrightarrow{[1]+(-1)[2]} \begin{bmatrix} 1 & 0 & | & 140 \\ 0 & 1 & | & 160 \end{bmatrix}$$

$x = 140$
(b)

33. Let x = adults, y = children
$\begin{cases} x + y = 600 \\ 5.5x + 2.5y = 1911 \end{cases}$

$\begin{bmatrix} 1 & 1 & | & 600 \\ 5.5 & 2.5 & | & 1911 \end{bmatrix}$

$$\xrightarrow{[2]+(-5.5)[1]} \begin{bmatrix} 1 & 1 & | & 600 \\ 0 & -3 & | & -1389 \end{bmatrix}$$

$$\xrightarrow{-\frac{1}{3}[3]} \begin{bmatrix} 1 & 1 & | & 600 \\ 0 & 1 & | & 463 \end{bmatrix}$$

$$\xrightarrow{[1]+(-1)[2]} \begin{bmatrix} 1 & 0 & | & 137 \\ 0 & 1 & | & 463 \end{bmatrix}$$

137 adults, 463 children

35. $\begin{cases} x + y + z = 100,000 \\ .08x + .07y + .1z = 8000 \\ x + y - 3z = 0 \end{cases}$

$$\begin{bmatrix} 1 & 1 & 1 & | & 100,000 \\ .08 & .07 & .1 & | & 8000 \\ 1 & 1 & -3 & | & 0 \end{bmatrix}$$

$\xrightarrow{[2]+(-.08)[1]}$ $\begin{bmatrix} 1 & 1 & 1 & | & 100,000 \\ 0 & -.01 & .02 & | & 0 \\ 1 & 1 & -3 & | & 0 \end{bmatrix}$

$\xrightarrow{[3]+(-1)[1]}$ $\begin{bmatrix} 1 & 1 & 1 & | & 100,000 \\ 0 & -.01 & .02 & | & 0 \\ 0 & 0 & -4 & | & -100,000 \end{bmatrix}$

$\xrightarrow{(-100)[2]}$ $\begin{bmatrix} 1 & 1 & 1 & | & 100,000 \\ 0 & 1 & -2 & | & 0 \\ 0 & 0 & -4 & | & -100,000 \end{bmatrix}$

$\xrightarrow{[1]+(-1)[2]}$ $\begin{bmatrix} 1 & 0 & 3 & | & 100,000 \\ 0 & 1 & -2 & | & 0 \\ 0 & 0 & -4 & | & -100,000 \end{bmatrix}$

$\xrightarrow{\left(-\frac{1}{4}\right)[3]}$ $\begin{bmatrix} 1 & 0 & 3 & | & 100,000 \\ 0 & 1 & -2 & | & 0 \\ 0 & 0 & 1 & | & 25,000 \end{bmatrix}$

$\xrightarrow{[1]+(-3)[3]}$ $\begin{bmatrix} 1 & 0 & 0 & | & 25,000 \\ 0 & 1 & -2 & | & 0 \\ 0 & 0 & 1 & | & 25,000 \end{bmatrix}$

$\xrightarrow{[2]+2[3]}$ $\begin{bmatrix} 1 & 0 & 0 & | & 25,000 \\ 0 & 1 & 0 & | & 50,000 \\ 0 & 0 & 1 & | & 25,000 \end{bmatrix}$

$x = \$25,000,\ y = \$50,000,\ z = \$25,000$

37.
```
[A]
    [[1  2  3]
     [-5 4  1]]
*row+(5,[A],1,2)
    [[1  2  3 ]
     [0 14 16]]
```

39.
```
[A]
    [[1 -.5 3]
     [0  1  4]]
rowSwap([A],1,2)
    [[0  1  4]
     [1 -.5 3]]
```

41. $\begin{bmatrix} 2 & 2 & 2 & 4 & | & -12 \\ 0 & 1 & 1 & 1 & | & -5 \\ 0 & 0 & 1 & 2 & | & -6 \\ 1 & 1 & 1 & 4 & | & -14 \end{bmatrix}$

$\xrightarrow{\frac{1}{2}[1]}$ $\begin{bmatrix} 1 & 1 & 1 & 2 & | & -6 \\ 0 & 1 & 1 & 1 & | & -5 \\ 0 & 0 & 1 & 2 & | & -6 \\ 1 & 1 & 1 & 4 & | & -14 \end{bmatrix}$

$\xrightarrow{[4]+(-1)[1]}$ $\begin{bmatrix} 1 & 1 & 1 & 2 & | & -6 \\ 0 & 1 & 1 & 1 & | & -5 \\ 0 & 0 & 1 & 2 & | & -6 \\ 0 & 0 & 0 & 2 & | & -8 \end{bmatrix}$

$\xrightarrow{[1]+(-1)[2]}$ $\begin{bmatrix} 1 & 0 & 0 & 1 & | & -1 \\ 0 & 1 & 1 & 1 & | & -5 \\ 0 & 0 & 1 & 2 & | & -6 \\ 0 & 0 & 0 & 2 & | & -8 \end{bmatrix}$

$\xrightarrow{[2]+(-1)[3]}$ $\begin{bmatrix} 1 & 0 & 0 & 1 & | & -1 \\ 0 & 1 & 0 & -1 & | & 1 \\ 0 & 0 & 1 & 2 & | & -6 \\ 0 & 0 & 0 & 2 & | & -8 \end{bmatrix}$

$\xrightarrow{\frac{1}{2}[4]}$ $\begin{bmatrix} 1 & 0 & 0 & 1 & | & -1 \\ 0 & 1 & 0 & -1 & | & 1 \\ 0 & 0 & 1 & 2 & | & -6 \\ 0 & 0 & 0 & 1 & | & -4 \end{bmatrix}$

$\xrightarrow{[1]+(-1)[4]}$ $\begin{bmatrix} 1 & 0 & 0 & 0 & | & 3 \\ 0 & 1 & 0 & -1 & | & 1 \\ 0 & 0 & 1 & 2 & | & -6 \\ 0 & 0 & 0 & 1 & | & -4 \end{bmatrix}$

$\xrightarrow{[2]+1[4]}$ $\begin{bmatrix} 1 & 0 & 0 & 0 & | & 3 \\ 0 & 1 & 0 & 0 & | & -3 \\ 0 & 0 & 1 & 2 & | & -6 \\ 0 & 0 & 0 & 1 & | & -4 \end{bmatrix}$

$$\xrightarrow{[3]+(-2)[4]} \begin{bmatrix} 1 & 0 & 0 & 0 & | & 3 \\ 0 & 1 & 0 & 0 & | & -3 \\ 0 & 0 & 1 & 0 & | & 2 \\ 0 & 0 & 0 & 1 & | & -4 \end{bmatrix}$$

$x = 3, y = -3, z = 2, w = -4$

Exercises 2

1. $\begin{bmatrix} 2 & -4 & 6 \\ 3 & 7 & 1 \end{bmatrix} \xrightarrow[{[2]+(-3)[1]}]{\frac{1}{2}[1]} \begin{bmatrix} 1 & -2 & 3 \\ 0 & 13 & -8 \end{bmatrix}$

3. $\begin{bmatrix} 7 & 1 & 4 & 5 \\ -1 & 1 & 2 & 6 \\ 4 & 0 & 2 & 3 \end{bmatrix}$

$$\xrightarrow[{[3]+(-2)[2]}]{\substack{\frac{1}{2}[2] \\ [1]+(-4)[2]}} \begin{bmatrix} 9 & -1 & 0 & -7 \\ -\frac{1}{2} & \frac{1}{2} & 1 & 3 \\ 5 & -1 & 0 & -3 \end{bmatrix}$$

5. $\begin{bmatrix} 2 & 3 \\ 6 & 0 \\ 1 & 5 \end{bmatrix} \xrightarrow[{[3]+(-1)[1]}]{\substack{\frac{1}{2}[1] \\ [2]+(-6)[1]}} \begin{bmatrix} 1 & \frac{3}{2} \\ 0 & -9 \\ 0 & \frac{7}{2} \end{bmatrix}$

7. $\begin{bmatrix} 4 & 3 & 0 \\ \frac{2}{3} & 0 & -2 \\ 1 & 3 & 6 \end{bmatrix} \xrightarrow[{[2]+2[3]}]{\frac{1}{6}[3]} \begin{bmatrix} 4 & 3 & 0 \\ 1 & 1 & 0 \\ \frac{1}{6} & \frac{1}{2} & 1 \end{bmatrix}$

9. $\begin{bmatrix} 2 & -4 & | & 6 \\ -1 & 2 & | & -3 \end{bmatrix}$

$\begin{bmatrix} 1 & -2 & | & 3 \\ 0 & 0 & | & 0 \end{bmatrix}$

$\begin{cases} x - 2y = 3 \\ \quad\quad 0 = 0 \end{cases}$

y = any value, $x = 2y + 3$

11. $\begin{bmatrix} 1 & 2 & | & 5 \\ 3 & -1 & | & 1 \\ -1 & 3 & | & 5 \end{bmatrix}$

$\begin{bmatrix} 1 & 2 & | & 5 \\ 0 & -7 & | & -14 \\ 0 & 5 & | & 10 \end{bmatrix}$

$\begin{bmatrix} 1 & 0 & | & 1 \\ 0 & 1 & | & 2 \\ 0 & 0 & | & 0 \end{bmatrix}$

$x = 1, y = 2$

13. $\begin{bmatrix} 1 & -1 & 3 & | & 3 \\ -2 & 3 & -11 & | & -4 \\ 1 & -2 & 8 & | & 6 \end{bmatrix}$

$\begin{bmatrix} 1 & -1 & 3 & | & 3 \\ 0 & 1 & -5 & | & 2 \\ 0 & -1 & 5 & | & 3 \end{bmatrix}$

$\begin{bmatrix} 1 & 0 & -2 & | & 5 \\ 0 & 1 & -5 & | & 2 \\ 0 & 0 & 0 & | & 5 \end{bmatrix}$

$\begin{cases} x - 2z = 5 \\ y - 5z = 2 \\ \quad\quad 0 = 5 \end{cases}$

No solution

15. $\begin{bmatrix} 1 & 1 & 1 & | & -1 \\ 2 & 3 & 2 & | & 3 \\ 2 & 1 & 2 & | & -7 \end{bmatrix}$

$\begin{bmatrix} 1 & 1 & 1 & | & -1 \\ 0 & 1 & 0 & | & 5 \\ 0 & -1 & 0 & | & -5 \end{bmatrix}$

$\begin{bmatrix} 1 & 0 & 1 & | & -6 \\ 0 & 1 & 0 & | & 5 \\ 0 & 0 & 0 & | & 0 \end{bmatrix}$

$$\begin{cases} x + z = -6 \\ \quad\quad y = 5 \\ \quad\quad 0 = 0 \end{cases}$$

$z = $ any value, $x = -z - 6$, $y = 5$

17. $\begin{bmatrix} \underline{1} & 2 & 3 & | & 4 \\ 5 & 6 & 7 & | & 8 \\ 1 & 2 & 3 & | & 5 \end{bmatrix}$

$\begin{bmatrix} 1 & 2 & 3 & | & 4 \\ 0 & \underline{-4} & -8 & | & -12 \\ 0 & 0 & 0 & | & 1 \end{bmatrix}$

$\begin{bmatrix} 1 & 0 & -1 & | & -2 \\ 0 & 1 & 2 & | & 3 \\ 0 & 0 & 0 & | & 1 \end{bmatrix}$

$$\begin{cases} x - z = -2 \\ y + 2z = 3 \\ \quad\quad 0 = 1 \end{cases}$$

No solution

19. $\begin{bmatrix} \underline{1} & 1 & -2 & 2 & | & 5 \\ 2 & 1 & -4 & 1 & | & 5 \\ 3 & 4 & -6 & 9 & | & 20 \\ 4 & 4 & -8 & 8 & | & 20 \end{bmatrix}$

$\begin{bmatrix} 1 & 1 & -2 & 2 & | & 5 \\ 0 & \underline{-1} & 0 & -3 & | & -5 \\ 0 & 1 & 0 & 3 & | & 5 \\ 0 & 0 & 0 & 0 & | & 0 \end{bmatrix}$

$\begin{bmatrix} 1 & 0 & -2 & -1 & | & 0 \\ 0 & 1 & 0 & 3 & | & 5 \\ 0 & 0 & 0 & 0 & | & 0 \\ 0 & 0 & 0 & 0 & | & 0 \end{bmatrix}$

$$\begin{cases} x - 2z - w = 0 \\ \quad\quad y + 3w = 5 \\ \quad\quad\quad 0 = 0 \\ \quad\quad\quad 0 = 0 \end{cases}$$

$z = $ any value, $w = $ any value, $x = 2z + w$, $y = -3w + 5$

21. $\begin{bmatrix} \underline{6} & -4 & | & 2 \\ -3 & 3 & | & 6 \\ 5 & 2 & | & 39 \end{bmatrix}$

$\begin{bmatrix} 1 & -\frac{2}{3} & | & \frac{1}{3} \\ 0 & 1 & | & 7 \\ 0 & \frac{16}{3} & | & \frac{112}{3} \end{bmatrix}$

$\begin{bmatrix} 1 & 0 & | & 5 \\ 0 & 1 & | & 7 \\ 0 & 0 & | & 0 \end{bmatrix}$

$x = 5$, $y = 7$

23. $\begin{bmatrix} \underline{1} & 2 & 1 & | & 5 \\ 0 & 1 & 3 & | & 9 \end{bmatrix}$

$\begin{bmatrix} 1 & 2 & 1 & | & 5 \\ 0 & \underline{1} & 3 & | & 9 \end{bmatrix}$

$\begin{bmatrix} 1 & 0 & -5 & | & -13 \\ 0 & 1 & 3 & | & 9 \end{bmatrix}$

$$\begin{cases} x - 5z = -13 \\ y + 3z = 9 \end{cases}$$

$z = $ any value, $x = 5z - 13$, $y = -3z + 9$
Possible answers: $z = 0$, $x = -13$, $y = 9$;
$z = 1$, $x = -8$, $y = 6$; $z = 2$, $x = -3$, $y = 3$

25. $\begin{bmatrix} \underline{1} & 7 & -3 & | & 8 \\ 0 & 0 & 1 & | & 5 \end{bmatrix}$

$\begin{bmatrix} 1 & 7 & -3 & | & 8 \\ 0 & 0 & \underline{1} & | & 5 \end{bmatrix}$

$\begin{bmatrix} 1 & 7 & 0 & | & 23 \\ 0 & 0 & 1 & | & 5 \end{bmatrix}$

$$\begin{cases} x + 7y = 23 \\ \quad\quad z = 5 \end{cases}$$

$y = $ any value, $x = -7y + 23$, $z = 5$
Possible answers: $y = 0$, $x = 23$, $z = 5$;
$y = 1$, $x = 16$, $z = 5$; $y = 2$, $x = 9$, $z = 5$

27. Let $x = $ food 1, $y = $ food 2, and $z = $ food 3.
$$\begin{cases} 2x + 4y + 6z = 1000 \\ 3x + 7y + 10z = 1600 \\ 5x + 9y + 14z = 2400 \end{cases}$$

$$\begin{bmatrix} \underline{2} & 4 & 6 & | & 1000 \\ 3 & 7 & 10 & | & 1600 \\ 5 & 9 & 14 & | & 2400 \end{bmatrix}$$

$$\begin{bmatrix} 1 & 2 & 3 & | & 500 \\ 0 & \underline{1} & 1 & | & 100 \\ 0 & -1 & -1 & | & -100 \end{bmatrix}$$

$$\begin{bmatrix} 1 & 0 & 1 & | & 300 \\ 0 & 1 & 1 & | & 100 \\ 0 & 0 & 0 & | & 0 \end{bmatrix}$$

$$\begin{cases} x + z = 300 \\ y + z = 100 \\ \quad\ 0 = 0 \end{cases}$$

$z =$ any amount, $x = 300 - z$, $y = 100 - z$
$(0 \le z \le 100)$

29. $\begin{cases} g + b + f = 8 \times 12 \\ g + b - 15f = 0 \\ 3g + 3b + 5f = 300 \end{cases}$

$$\begin{bmatrix} \underline{1} & 1 & 1 & | & 96 \\ 1 & 1 & -15 & | & 0 \\ 3 & 3 & 5 & | & 300 \end{bmatrix}$$

$$\begin{bmatrix} 1 & 1 & 1 & | & 96 \\ 0 & 0 & \underline{-16} & | & -96 \\ 0 & 0 & 2 & | & 12 \end{bmatrix}$$

$$\begin{bmatrix} 1 & 1 & 0 & | & 90 \\ 0 & 0 & 1 & | & 6 \\ 0 & 0 & 0 & | & 0 \end{bmatrix}$$

$$\begin{cases} g + b = 90 \\ \quad\ f = 6 \\ \quad\ 0 = 0 \end{cases}$$

6 floral squares, the other 90 any mix of solid green and blue.

31. $\begin{bmatrix} \underline{1} & 1 & 1 & | & 14 \\ 1 & -1 & 2 & | & 15 \\ 1 & 2 & 3 & | & 36 \end{bmatrix}$

$$\begin{bmatrix} 1 & 1 & 1 & | & 14 \\ 0 & \underline{-2} & 1 & | & 1 \\ 0 & 1 & 2 & | & 22 \end{bmatrix}$$

$$\begin{bmatrix} 1 & 0 & \frac{3}{2} & | & \frac{29}{2} \\ 0 & 1 & -\frac{1}{2} & | & -\frac{1}{2} \\ 0 & 0 & \frac{5}{2} & | & \frac{45}{2} \end{bmatrix}$$

$$\begin{bmatrix} 1 & 0 & 0 & | & 1 \\ 0 & 1 & 0 & | & 4 \\ 0 & 0 & 1 & | & 9 \end{bmatrix}$$

$$\begin{cases} x^2 = 1 \\ y^2 = 4 \\ z^2 = 9 \end{cases}$$

$x = \pm 1$, $y = \pm 2$, $z = \pm 3$

33. $\begin{bmatrix} 2 & -3 & | & 4 \\ -6 & 9 & | & k \end{bmatrix}$

$$\begin{bmatrix} 1 & -\frac{3}{2} & | & 2 \\ 0 & 0 & | & 12 + k \end{bmatrix}$$

No solution if $0 \ne 12 + k$, which happens when $k \ne -12$.
Infinitely many if $0 = 12 + k$, which happens when $k = -12$.

35.

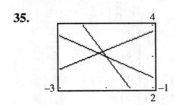

No solution

37.

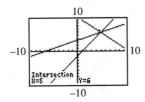

39. $\begin{bmatrix} 2 & 3 & 6 & | & 4 \\ 4 & 7 & 0 & | & 2 \\ 3 & 5 & 3 & | & 3 \end{bmatrix}$

$\begin{bmatrix} 1 & \frac{3}{2} & 3 & | & 2 \\ 0 & 1 & -12 & | & -6 \\ 0 & \frac{1}{2} & -6 & | & -3 \end{bmatrix}$

$\begin{bmatrix} 1 & 0 & 21 & | & 11 \\ 0 & 1 & -12 & | & -6 \\ 0 & 0 & 0 & | & 0 \end{bmatrix}$

$\begin{cases} x + 21z = 11 \\ y - 12z = -6 \\ \quad\quad 0 = 0 \end{cases}$

z = any value, $x = -21z + 11$, $y = 12z - 6$

41. $\begin{bmatrix} 8 & 3 & -2 & | & 5 \\ 12 & 5 & 2 & | & 3 \\ 5 & 2 & 0 & | & 7 \end{bmatrix}$

$\begin{bmatrix} 1 & \frac{3}{8} & -\frac{1}{4} & | & \frac{5}{8} \\ 0 & \frac{1}{2} & 5 & | & -\frac{9}{2} \\ 0 & \frac{1}{8} & \frac{5}{4} & | & \frac{31}{8} \end{bmatrix}$

$\begin{bmatrix} 1 & 0 & -4 & | & 4 \\ 0 & 1 & 10 & | & -9 \\ 0 & 0 & 0 & | & 5 \end{bmatrix}$

$\begin{cases} x - 4z = 4 \\ y + 10z = -9 \\ \quad\quad 0 = 5 \end{cases}$

No solution

Exercises 3

1. 2×3

3. 1×3, row matrix

5. 1×1, square matrix

7. $\begin{bmatrix} 4+5 & -2+5 \\ 3+4 & 0+(-1) \end{bmatrix} = \begin{bmatrix} 9 & 3 \\ 7 & -1 \end{bmatrix}$

9. $\begin{bmatrix} 2-1 & 8-5 \\ \frac{4}{3}-\frac{1}{3} & 4-2 \\ 1-(-3) & -2-0 \end{bmatrix} = \begin{bmatrix} 1 & 3 \\ 1 & 2 \\ 4 & -2 \end{bmatrix}$

11. $[5 \cdot 1 + 3 \cdot 2] = [11]$

13. $\left[6 \cdot \frac{1}{2} + 1(-3) + 5 \cdot 2 \right] = [10]$

15. Yes, columns of A = rows of B; 3×5

17. No, columns of $A \neq$ rows of B

19. Yes, columns of A = rows of B; 3×1

21. $\begin{bmatrix} 3 \cdot 1 + 1 \cdot 3 & 3 \cdot 4 + 1 \cdot 5 \\ 0 \cdot 1 + 2 \cdot 3 & 0 \cdot 4 + 2 \cdot 5 \end{bmatrix} = \begin{bmatrix} 6 & 17 \\ 6 & 10 \end{bmatrix}$

23. $\begin{bmatrix} 4 \cdot 5 + 1 \cdot 1 + 0 \cdot 2 \\ -2 \cdot 5 + 0 \cdot 1 + 3 \cdot 2 \\ 1 \cdot 5 + 5 \cdot 1 + (-1)2 \end{bmatrix} = \begin{bmatrix} 21 \\ -4 \\ 8 \end{bmatrix}$

25. Multiplication by identity matrix:
$\begin{bmatrix} 5 & 6 \\ 7 & 8 \end{bmatrix}$

27. $\begin{bmatrix} (.6)(.6)+(.3)(.4) & (.6)(.3)+(.3)(.7) \\ (.4)(.6)+(.7)(.4) & (.4)(.3)+(.7)(.7) \end{bmatrix} = \begin{bmatrix} .48 & .39 \\ .52 & .61 \end{bmatrix}$

29. $\begin{bmatrix} 2\cdot 4+(-1)3+4\cdot 5 & 2\cdot 8+(-1)(-1)+4\cdot 0 & 2\cdot 0+(-1)2+4\cdot 1 \\ 0\cdot 4+1\cdot 3+0\cdot 5 & 0\cdot 8+1(-1)+0\cdot 0 & 0\cdot 0+1\cdot 2+0\cdot 1 \\ \frac{1}{2}\cdot 4+3\cdot 3+(-2)5 & \frac{1}{2}\cdot 8+3(-1)+(-2)0 & \frac{1}{2}\cdot 0+3\cdot 2+(-2)\cdot 1 \end{bmatrix} = \begin{bmatrix} 25 & 17 & 2 \\ 3 & -1 & 2 \\ 1 & 1 & 4 \end{bmatrix}$

31. $\begin{cases} 2x+3y=6 \\ 4x+5y=7 \end{cases}$

33. $\begin{cases} x+2y+3z=10 \\ 4x+5y+6z=11 \\ 7x+8y+9z=12 \end{cases}$

35. $\begin{bmatrix} 3 & 2 \\ 7 & -1 \end{bmatrix}\begin{bmatrix} x \\ y \end{bmatrix} = \begin{bmatrix} -1 \\ 2 \end{bmatrix}$

37. $\begin{bmatrix} 1 & -2 & 3 \\ 0 & 1 & 1 \\ 0 & 0 & 1 \end{bmatrix}\begin{bmatrix} x \\ y \\ z \end{bmatrix} = \begin{bmatrix} 5 \\ 6 \\ 2 \end{bmatrix}$

39. $\left(\begin{bmatrix} 1 & 2 \\ 0 & 3 \end{bmatrix} + \begin{bmatrix} 3 & -2 \\ 4 & 5 \end{bmatrix} \right)\begin{bmatrix} 1 & 6 \\ 2 & 0 \end{bmatrix} = \begin{bmatrix} 4 & 0 \\ 4 & 8 \end{bmatrix}\begin{bmatrix} 1 & 6 \\ 2 & 0 \end{bmatrix} = \begin{bmatrix} 4 & 24 \\ 20 & 24 \end{bmatrix}$

$\begin{bmatrix} 1 & 2 \\ 0 & 3 \end{bmatrix}\begin{bmatrix} 1 & 6 \\ 2 & 0 \end{bmatrix} + \begin{bmatrix} 3 & -2 \\ 4 & 5 \end{bmatrix}\begin{bmatrix} 1 & 6 \\ 2 & 0 \end{bmatrix} = \begin{bmatrix} 5 & 6 \\ 6 & 0 \end{bmatrix} + \begin{bmatrix} -1 & 18 \\ 14 & 24 \end{bmatrix} = \begin{bmatrix} 4 & 24 \\ 20 & 24 \end{bmatrix}$

41. $\begin{bmatrix} 3\cdot 1+(-1)2 & 3\cdot 2+(-1)6 \\ -1\cdot 1+\frac{1}{2}\cdot 2 & -1\cdot 2+\frac{1}{2}\cdot 6 \end{bmatrix} = \begin{bmatrix} 1 & 0 \\ 0 & 1 \end{bmatrix}$

43. a. $\begin{bmatrix} 6 & 8 & 2 \\ 2 & 5 & 3 \end{bmatrix}\begin{bmatrix} 20 \\ 15 \\ 50 \end{bmatrix} = \begin{bmatrix} 340 \\ 265 \end{bmatrix}$

    b. Mike's clothes cost \$340; Don's clothes cost \$265.

45. a.

$$\begin{bmatrix} .25 & .35 & .30 & .10 & 0 \\ .10 & .20 & .40 & .20 & .10 \\ .05 & .10 & .20 & .40 & .25 \end{bmatrix} \begin{bmatrix} 4 \\ 3 \\ 2 \\ 1 \\ 0 \end{bmatrix} = \begin{bmatrix} 2.75 \\ 2.00 \\ 1.30 \end{bmatrix}$$

I: 2.75; II: 2, III: 1.3

b. $[240 \quad 120 \quad 40] \begin{bmatrix} .25 & .35 & .30 & .10 & 0 \\ .10 & .20 & .40 & .20 & .10 \\ .05 & .10 & .20 & .40 & .25 \end{bmatrix} = [74 \quad 112 \quad 128 \quad 64 \quad 22]$

A: 74, B: 112, C: 128, D: 64, F: 22

47. $[6000 \quad 8000 \quad 4000] \begin{bmatrix} .65 & .35 \\ .55 & .45 \\ .45 & .55 \end{bmatrix} = [10,100 \quad 7900]$

10,100 voting Democratic, 7900 voting Republican

49. $\begin{bmatrix} 50 & 20 & 10 \\ 30 & 30 & 15 \\ 20 & 20 & 5 \end{bmatrix} \begin{bmatrix} 10 \\ 15 \\ 20 \end{bmatrix} = \begin{bmatrix} 1000 \\ 1050 \\ 600 \end{bmatrix}$

Carpenters: \$1000, bricklayers: \$1050, plumbers: \$600

51. a. $BN = [162 \quad 150 \quad 143]$, number of units of each nutrient consumed at breakfast

b. $LN = [186 \quad 200 \quad 239]$, number of units of each nutrient consumed at lunch

c. $DN = [288 \quad 300 \quad 344]$, number of units of each nutrient consumed at dinner

d. $B + L + D = [5 \quad 8]$, total number of ounces of each food that Mikey eats during a day

e. $(B + L + D)N = [636 \quad 650 \quad 726]$, number of units of each nutrient consumed per day

53. a. $AP = \begin{bmatrix} 720 \\ 646 \end{bmatrix}$, the average amount taken in daily by the pool and the weight room

b. \$720

55. Swap roles of x and z: $x = 3$, $y = 4$, $z = 5$

57. $(A + B) - A = \begin{bmatrix} 3 & -2 & 1 \\ -5 & 6 & 7 \end{bmatrix}$

59. $AB = \begin{bmatrix} 27.9 & 130.6 & -69.88 \\ 106.75 & -149.44 & 26.1 \\ -47.5 & 336.2 & -18.7 \end{bmatrix}$

61. $A(B + C) = \begin{bmatrix} -69.14 & 147.9 & -43.26 \\ 158.05 & -3.69 & 33.46 \\ -176.1 & 259.5 & 59.3 \end{bmatrix}$

63.

```
[A](2,3)
              1.6
```

65. $A^2 = AA = \begin{bmatrix} 160.16 & -26.7 & 4 \\ 2.7 & 150.85 & -53 \\ 187.4 & -35.5 & 48.6 \end{bmatrix}$

67. They match.

Exercises 4

1. $\begin{bmatrix} 1 & -2 \\ -\frac{1}{2} & 2 \end{bmatrix}\begin{bmatrix} 4 \\ 1 \end{bmatrix} = \begin{bmatrix} 2 \\ 0 \end{bmatrix}$

$x = 2, y = 0$

3. $\Delta = 7 \cdot 1 - 3 \cdot 2 = 1$

$\begin{bmatrix} \frac{1}{1} & -\frac{2}{1} \\ -\frac{3}{1} & \frac{7}{1} \end{bmatrix} = \begin{bmatrix} 1 & -2 \\ -3 & 7 \end{bmatrix}$

5. $\Delta = 6 \cdot 2 - 5 \cdot 2 = 2$

$\begin{bmatrix} \frac{2}{2} & -\frac{2}{2} \\ -\frac{5}{2} & \frac{6}{2} \end{bmatrix} = \begin{bmatrix} 1 & -1 \\ -\frac{5}{2} & 3 \end{bmatrix}$

7. $\Delta = (.7)(.8) - (.3)(.2) = .5$

$\begin{bmatrix} \frac{.8}{.5} & -\frac{.2}{.5} \\ -\frac{.3}{.5} & \frac{.7}{.5} \end{bmatrix} = \begin{bmatrix} 1.6 & -.4 \\ -.6 & 1.4 \end{bmatrix}$

9. For a 1×1 matrix $[a]$ $(a \neq 0)$, $[a]^{-1} = \left[\frac{1}{a}\right]$.

$\left[\frac{1}{3}\right]$

11. $\begin{bmatrix} 1 & 2 \\ 2 & 6 \end{bmatrix}^{-1}\begin{bmatrix} 3 \\ 5 \end{bmatrix} = \begin{bmatrix} 3 & -1 \\ -1 & \frac{1}{2} \end{bmatrix}\begin{bmatrix} 3 \\ 5 \end{bmatrix} = \begin{bmatrix} 4 \\ -\frac{1}{2} \end{bmatrix}$

$x = 4, \; y = -\frac{1}{2}$

13. $\begin{bmatrix} \frac{1}{2} & 2 \\ 3 & 16 \end{bmatrix}^{-1}\begin{bmatrix} 4 \\ 0 \end{bmatrix} = \begin{bmatrix} 8 & -1 \\ -\frac{3}{2} & \frac{1}{4} \end{bmatrix}\begin{bmatrix} 4 \\ 0 \end{bmatrix} = \begin{bmatrix} 32 \\ -6 \end{bmatrix}$

$x = 32, y = -6$

15. a. $\begin{bmatrix} .8 & .3 \\ .2 & .7 \end{bmatrix}\begin{bmatrix} x \\ y \end{bmatrix} = \begin{bmatrix} m \\ s \end{bmatrix}$

b. $\begin{bmatrix} x \\ y \end{bmatrix} = \begin{bmatrix} .8 & .3 \\ .2 & .7 \end{bmatrix}^{-1}\begin{bmatrix} m \\ s \end{bmatrix}$

$= \begin{bmatrix} 1.4 & -.6 \\ -.4 & 1.6 \end{bmatrix}\begin{bmatrix} m \\ s \end{bmatrix}$

c. $\begin{bmatrix} 1.4 & -.6 \\ -.4 & 1.6 \end{bmatrix}\begin{bmatrix} 100,000 \\ 50,000 \end{bmatrix} = \begin{bmatrix} 110,000 \\ 40,000 \end{bmatrix}$

110,000 married; 40,000 single

d. $\begin{bmatrix} 1.4 & -.6 \\ -.4 & 1.6 \end{bmatrix}\begin{bmatrix} 110,000 \\ 40,000 \end{bmatrix} = \begin{bmatrix} 130,000 \\ 20,000 \end{bmatrix}$

130,000 married; 20,000 single

17. a. $\begin{bmatrix} .7 & .1 \\ .3 & .9 \end{bmatrix}\begin{bmatrix} x \\ y \end{bmatrix} = \begin{bmatrix} u \\ v \end{bmatrix}$

b. $\begin{bmatrix} x \\ y \end{bmatrix} = \begin{bmatrix} .7 & .1 \\ .3 & .9 \end{bmatrix}^{-1}\begin{bmatrix} u \\ v \end{bmatrix} = \begin{bmatrix} \frac{3}{2} & -\frac{1}{6} \\ -\frac{1}{2} & \frac{7}{6} \end{bmatrix}\begin{bmatrix} u \\ v \end{bmatrix}$

c. $\begin{bmatrix} \frac{3}{2} & -\frac{1}{6} \\ -\frac{1}{2} & \frac{7}{6} \end{bmatrix}\begin{bmatrix} 6000 \\ 3000 \end{bmatrix} = \begin{bmatrix} 8500 \\ 500 \end{bmatrix}$

$\begin{bmatrix} .7 & .1 \\ .3 & .9 \end{bmatrix}\begin{bmatrix} 6000 \\ 3000 \end{bmatrix} = \begin{bmatrix} 4500 \\ 4500 \end{bmatrix}$

8500; 4500

19. $\begin{bmatrix} 5 & -2 & -2 \\ -1 & 1 & 0 \\ -1 & 0 & 1 \end{bmatrix}\begin{bmatrix} 1 \\ -1 \\ -1 \end{bmatrix} = \begin{bmatrix} 9 \\ -2 \\ -2 \end{bmatrix}$

$x = 9, y = -2, z = -2$

21.
$$\begin{bmatrix} 1 & 0 & -2 & 0 \\ 0 & 1 & 0 & -5 \\ -4 & 0 & 9 & 0 \\ 0 & 2 & 1 & -9 \end{bmatrix} \begin{bmatrix} 1 \\ 0 \\ 0 \\ -1 \end{bmatrix} = \begin{bmatrix} 1 \\ 5 \\ -4 \\ 9 \end{bmatrix}$$
$x = 1, y = 5, z = -4, w = 9$

23. Suppose $\begin{bmatrix} 6 & 3 \\ 2 & 1 \end{bmatrix}^{-1} = \begin{bmatrix} s & t \\ u & v \end{bmatrix}$.

Then $\begin{bmatrix} 6s+3u & 6t+3v \\ 2s+u & 2t+v \end{bmatrix} = \begin{bmatrix} 1 & 0 \\ 0 & 1 \end{bmatrix}$.

Then $\dfrac{6s+3u}{3} = 2s+u = \dfrac{1}{3}$, which

contradicts $2s + u = 0$.

25. a. $\begin{cases} x + 2y = a \\ \quad .9x = b \end{cases}$

$$\begin{bmatrix} 1 & 2 \\ .9 & 0 \end{bmatrix} \begin{bmatrix} x \\ y \end{bmatrix} = \begin{bmatrix} a \\ b \end{bmatrix}$$

b. $\begin{bmatrix} 1 & 2 \\ .9 & 0 \end{bmatrix} \begin{bmatrix} 450,000 \\ 360,000 \end{bmatrix} = \begin{bmatrix} 1,170,000 \\ 405,000 \end{bmatrix}$

After 1 year:
1,170,000 in group I,
405,000 in group II

$\begin{bmatrix} 1 & 2 \\ .9 & 0 \end{bmatrix} \begin{bmatrix} 1,170,000 \\ 405,000 \end{bmatrix} = \begin{bmatrix} 1,980,000 \\ 1,053,000 \end{bmatrix}$

After 2 years:
1,980,000 in group I,
1,053,000 in group II

c. $\begin{bmatrix} 1 & 2 \\ .9 & 0 \end{bmatrix}^{-1} \begin{bmatrix} 810,000 \\ 630,000 \end{bmatrix} = \begin{bmatrix} 700,000 \\ 55,000 \end{bmatrix}$

700,000 in group I, 55,000 in group II

27.
```
[A]⁻¹
[[-.1369863014 …
 [.3424657534  …
Ans▶Frac
[[-10/73 75/292…
 [25/73  -5/292…
```

$$\begin{bmatrix} -\frac{10}{73} & \frac{75}{292} \\ \frac{25}{73} & -\frac{5}{292} \end{bmatrix}$$

29.
```
[[.1147743896 :…
 [.3431979296 :…
 [.0140654889 :…
Ans▶Frac
[[1020/8887 291…
 [3050/8887 860…
 [125/8887  618…
```

$$\begin{bmatrix} \frac{1020}{8887} & \frac{2910}{8887} & -\frac{500}{8887} \\ \frac{3050}{8887} & \frac{860}{8887} & \frac{1990}{8887} \\ \frac{125}{8887} & \frac{618}{8887} & \frac{810}{8887} \end{bmatrix}$$

31.
```
[A]⁻¹[B]▶Frac
        [[-4/5]
         [28/5]
         [5   ]]
```

$x = -\dfrac{4}{5}, \; y = \dfrac{28}{5}, \; z = 5$

33.
```
[A]⁻¹[B]▶Frac
          [[0]
           [2]
           [0]
           [2]]
```

$x = 0, y = 2, z = 0, w = 2$

35. With the message ERR:SINGULAR MAT

Exercises 5

1. $\begin{bmatrix} 7 & 3 & | & 1 & 0 \\ 5 & 2 & | & 0 & 1 \end{bmatrix}$

$\begin{bmatrix} 1 & \frac{3}{7} & | & \frac{1}{7} & 0 \\ 0 & -\frac{1}{7} & | & -\frac{5}{7} & 1 \end{bmatrix}$

$\begin{bmatrix} 1 & 0 & | & -2 & 3 \\ 0 & 1 & | & 5 & -7 \end{bmatrix}$

$\begin{bmatrix} -2 & 3 \\ 5 & -7 \end{bmatrix}$

3. $\begin{bmatrix} 10 & 12 & | & 1 & 0 \\ 3 & -4 & | & 0 & 1 \end{bmatrix}$

$\begin{bmatrix} 1 & \frac{6}{5} & | & \frac{1}{10} & 0 \\ 0 & -\frac{38}{5} & | & -\frac{3}{10} & 1 \end{bmatrix}$

$$\begin{bmatrix} 1 & 0 & | & \frac{1}{19} & \frac{3}{19} \\ 0 & 1 & | & \frac{3}{76} & -\frac{5}{38} \end{bmatrix}$$

$$\begin{bmatrix} \frac{1}{19} & \frac{3}{19} \\ \frac{3}{76} & -\frac{5}{36} \end{bmatrix}$$

5. $\begin{bmatrix} \underline{2} & -4 & | & 1 & 0 \\ -1 & 2 & | & 0 & 1 \end{bmatrix}$

$$\begin{bmatrix} 1 & -2 & | & \frac{1}{2} & 0 \\ 0 & 0 & | & \frac{1}{2} & 1 \end{bmatrix}$$

No inverse

7. $\begin{bmatrix} \underline{1} & 2 & -2 & | & 1 & 0 & 0 \\ 1 & 1 & 1 & | & 0 & 1 & 0 \\ 0 & 0 & 1 & | & 0 & 0 & 1 \end{bmatrix}$

$$\begin{bmatrix} 1 & 2 & -2 & | & 1 & 0 & 0 \\ 0 & \underline{-1} & 3 & | & -1 & 1 & 0 \\ 0 & 0 & 1 & | & 0 & 0 & 1 \end{bmatrix}$$

$$\begin{bmatrix} 1 & 0 & 4 & | & -1 & 2 & 0 \\ 0 & 1 & -3 & | & 1 & -1 & 0 \\ 0 & 0 & \underline{1} & | & 0 & 0 & 1 \end{bmatrix}$$

$$\begin{bmatrix} 1 & 0 & 0 & | & -1 & 2 & -4 \\ 0 & 1 & 0 & | & 1 & -1 & 3 \\ 0 & 0 & 1 & | & 0 & 0 & 1 \end{bmatrix}$$

$$\begin{bmatrix} -1 & 2 & -4 \\ 1 & -1 & 3 \\ 0 & 0 & 1 \end{bmatrix}$$

9. $\begin{bmatrix} \underline{-2} & 5 & 2 & | & 1 & 0 & 0 \\ 1 & -3 & -1 & | & 0 & 1 & 0 \\ -1 & 2 & 1 & | & 0 & 0 & 1 \end{bmatrix}$

$$\begin{bmatrix} 1 & -\frac{5}{2} & -1 & | & -\frac{1}{2} & 0 & 0 \\ 0 & -\frac{1}{2} & 0 & | & \frac{1}{2} & 1 & 0 \\ 0 & -\frac{1}{2} & 0 & | & -\frac{1}{2} & 0 & 1 \end{bmatrix}$$

$$\begin{bmatrix} 1 & 0 & -1 & | & -3 & -5 & 0 \\ 0 & 1 & 0 & | & -1 & -2 & 0 \\ 0 & 0 & 0 & | & -1 & -1 & 1 \end{bmatrix}$$

No inverse

11. $\begin{bmatrix} \underline{1} & 6 & 0 & 0 & | & 1 & 0 & 0 & 0 \\ 1 & 5 & 0 & 0 & | & 0 & 1 & 0 & 0 \\ 0 & 0 & 4 & 2 & | & 0 & 0 & 1 & 0 \\ 0 & 0 & 50 & 2 & | & 0 & 0 & 0 & 1 \end{bmatrix}$

$$\begin{bmatrix} 1 & 6 & 0 & 0 & | & 1 & 0 & 0 & 0 \\ 0 & \underline{-1} & 0 & 0 & | & -1 & 1 & 0 & 0 \\ 0 & 0 & 4 & 2 & | & 0 & 0 & 1 & 0 \\ 0 & 0 & 50 & 2 & | & 0 & 0 & 0 & 1 \end{bmatrix}$$

$$\begin{bmatrix} 1 & 0 & 0 & 0 & | & -5 & 6 & 0 & 0 \\ 0 & 1 & 0 & 0 & | & 1 & -1 & 0 & 0 \\ 0 & 0 & \underline{4} & 2 & | & 0 & 0 & 1 & 0 \\ 0 & 0 & 50 & 2 & | & 0 & 0 & 0 & 1 \end{bmatrix}$$

$$\begin{bmatrix} 1 & 0 & 0 & 0 & | & -5 & 6 & 0 & 0 \\ 0 & 1 & 0 & 0 & | & 1 & -1 & 0 & 0 \\ 0 & 0 & 1 & \frac{1}{2} & | & 0 & 0 & \frac{1}{4} & 0 \\ 0 & 0 & 0 & \underline{-23} & | & 0 & 0 & -\frac{25}{2} & 1 \end{bmatrix}$$

$$\begin{bmatrix} 1 & 0 & 0 & 0 & | & -5 & 6 & 0 & 0 \\ 0 & 1 & 0 & 0 & | & 1 & -1 & 0 & 0 \\ 0 & 0 & 1 & 0 & | & 0 & 0 & -\frac{1}{46} & \frac{1}{46} \\ 0 & 0 & 0 & 1 & | & 0 & 0 & \frac{25}{46} & -\frac{1}{23} \end{bmatrix}$$

$$\begin{bmatrix} -5 & 6 & 0 & 0 \\ 1 & -1 & 0 & 0 \\ 0 & 0 & -\frac{1}{46} & \frac{1}{46} \\ 0 & 0 & \frac{25}{46} & -\frac{1}{23} \end{bmatrix}$$

13. Find the inverse of $\begin{bmatrix} 1 & 1 & 2 \\ 3 & 2 & 2 \\ 1 & 1 & 3 \end{bmatrix}$.

$$\begin{bmatrix} 1 & 1 & 2 & | & 1 & 0 & 0 \\ 3 & 2 & 2 & | & 0 & 1 & 0 \\ 1 & 1 & 3 & | & 0 & 0 & 1 \end{bmatrix}$$

$$\begin{bmatrix} 1 & 1 & 2 & | & 1 & 0 & 0 \\ 0 & -1 & -4 & | & -3 & 1 & 0 \\ 0 & 0 & 1 & | & -1 & 0 & 1 \end{bmatrix}$$

$$\left[\begin{array}{ccc|ccc} 1 & 0 & -2 & -2 & 1 & 0 \\ 0 & 1 & 4 & 3 & -1 & 0 \\ 0 & 0 & 1 & -1 & 0 & 1 \end{array}\right]$$

$$\left[\begin{array}{ccc|ccc} 1 & 0 & 0 & -4 & 1 & 2 \\ 0 & 1 & 0 & 7 & -1 & -4 \\ 0 & 0 & 1 & -1 & 0 & 1 \end{array}\right]$$

$$\left[\begin{array}{ccc} 1 & 1 & 2 \\ 3 & 2 & 2 \\ 1 & 1 & 3 \end{array}\right]^{-1} \left[\begin{array}{c} 3 \\ 4 \\ 5 \end{array}\right] = \left[\begin{array}{ccc} -4 & 1 & 2 \\ 7 & -1 & -4 \\ -1 & 0 & 1 \end{array}\right]\left[\begin{array}{c} 3 \\ 4 \\ 5 \end{array}\right] = \left[\begin{array}{c} 2 \\ -3 \\ 2 \end{array}\right]$$

$x = 2, y = -3, z = 2$

15. Find the inverse of $\left[\begin{array}{cccc} 1 & 0 & -2 & -2 \\ 0 & 1 & 0 & -5 \\ -4 & 0 & 9 & 9 \\ 0 & 2 & 1 & -8 \end{array}\right]$.

$$\left[\begin{array}{cccc|cccc} 1 & 0 & -2 & -2 & 1 & 0 & 0 & 0 \\ 0 & 1 & 0 & -5 & 0 & 1 & 0 & 0 \\ -4 & 0 & 9 & 9 & 0 & 0 & 1 & 0 \\ 0 & 2 & 1 & -8 & 0 & 0 & 0 & 1 \end{array}\right]$$

$$\left[\begin{array}{cccc|cccc} 1 & 0 & -2 & -2 & 1 & 0 & 0 & 0 \\ 0 & 1 & 0 & -5 & 0 & 1 & 0 & 0 \\ 0 & 0 & 1 & 1 & 4 & 0 & 1 & 0 \\ 0 & 2 & 1 & -8 & 0 & 0 & 0 & 1 \end{array}\right]$$

$$\left[\begin{array}{cccc|cccc} 1 & 0 & -2 & -2 & 1 & 0 & 0 & 0 \\ 0 & 1 & 0 & -5 & 0 & 1 & 0 & 0 \\ 0 & 0 & 1 & 1 & 4 & 0 & 1 & 0 \\ 0 & 0 & 1 & 2 & 0 & -2 & 0 & 1 \end{array}\right]$$

$$\left[\begin{array}{cccc|cccc} 1 & 0 & 0 & 0 & 9 & 0 & 2 & 0 \\ 0 & 1 & 0 & -5 & 0 & 1 & 0 & 0 \\ 0 & 0 & 1 & 1 & 4 & 0 & 1 & 0 \\ 0 & 0 & 0 & 1 & -4 & -2 & -1 & 1 \end{array}\right]$$

$$\left[\begin{array}{cccc|cccc} 1 & 0 & 0 & 0 & 9 & 0 & 2 & 0 \\ 0 & 1 & 0 & 0 & -20 & -9 & -5 & 5 \\ 0 & 0 & 1 & 0 & 8 & 2 & 2 & -1 \\ 0 & 0 & 0 & 1 & -4 & -2 & -1 & 1 \end{array}\right]$$

$$\left[\begin{array}{cccc} 1 & 0 & -2 & -2 \\ 0 & 1 & 0 & -5 \\ -4 & 0 & 9 & 9 \\ 0 & 2 & 1 & -8 \end{array}\right]^{-1}\left[\begin{array}{c} 0 \\ 1 \\ 2 \\ 3 \end{array}\right]$$

$$= \left[\begin{array}{cccc} 9 & 0 & 2 & 0 \\ -20 & -9 & -5 & 5 \\ 8 & 2 & 2 & -1 \\ -4 & 2 & -1 & 1 \end{array}\right]\left[\begin{array}{c} 0 \\ 1 \\ 2 \\ 3 \end{array}\right]$$

$$= \left[\begin{array}{c} 4 \\ -4 \\ 3 \\ -1 \end{array}\right]$$

$x = 4, y = -4, z = 3, w = -1$

17. Either no solution or infinitely many solutions

19. $A \cdot \left[\begin{array}{cc} 2 & 5 \\ 1 & 3 \end{array}\right] = \left[\begin{array}{cc} -1 & 0 \\ 4 & 2 \end{array}\right]$

$A = A \cdot \left[\begin{array}{cc} 2 & 5 \\ 1 & 3 \end{array}\right]\left[\begin{array}{cc} 2 & 5 \\ 1 & 3 \end{array}\right]^{-1}$

$= \left[\begin{array}{cc} -1 & 0 \\ 4 & 2 \end{array}\right]\left[\begin{array}{cc} 2 & 5 \\ 1 & 3 \end{array}\right]^{-1} = \left[\begin{array}{cc} -1 & 0 \\ 4 & 2 \end{array}\right]\left[\begin{array}{cc} 3 & -5 \\ -1 & 2 \end{array}\right]$

$= \left[\begin{array}{cc} -3 & 5 \\ 10 & -16 \end{array}\right]$

21.

```
[A]
    [[4 1 1 0]
     [7 2 0 1]]
rref([A])
    [[1 0 2 -1]
     [0 1 -7 4 ]]
```

$\left[\begin{array}{cc} 2 & -1 \\ -7 & 4 \end{array}\right]$

23.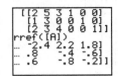

$$\begin{bmatrix} -2.4 & 2.2 & 1.8 \\ .8 & -.4 & -.6 \\ .6 & -.8 & -.2 \end{bmatrix}$$

Exercises 6

1. $D_{\text{new}} = \begin{bmatrix} 4 \\ 1.5 \\ 9 \end{bmatrix}$

$(I - A)^{-1} D_{\text{new}} = \begin{bmatrix} 1.01 & .20 & .50 \\ .02 & 1.05 & .23 \\ .01 & .09 & 1.08 \end{bmatrix} \begin{bmatrix} 4 \\ 1.5 \\ 9 \end{bmatrix}$

$= \begin{bmatrix} 8.84 \\ 3.725 \\ 9.895 \end{bmatrix}$

Coal: \$8.84 billion, steel: \$3.725 billion, electricity: \$9.895 billion

3. $D_{\text{new}} = \begin{bmatrix} 3 \\ 1 \\ 4 \end{bmatrix}$

$(I - A)^{-1} D_{\text{new}} = \begin{bmatrix} 1.04 & .02 & .10 \\ .21 & 1.01 & .02 \\ .11 & .02 & 1.02 \end{bmatrix} \begin{bmatrix} 3 \\ 1 \\ 4 \end{bmatrix}$

$= \begin{bmatrix} 3.54 \\ 1.72 \\ 4.43 \end{bmatrix}$

Computers: \$354 million, semiconductors: \$172 million

5.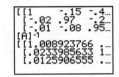

7. $A = \begin{bmatrix} .02 & .01 \\ .10 & .05 \end{bmatrix} \begin{matrix} P \\ I \end{matrix}$ with columns P I

$D = \begin{bmatrix} 930 \\ 465 \end{bmatrix}$

$(I - A)^{-1} D = \begin{bmatrix} 955 \\ 590 \end{bmatrix}$

Plastics: \$955,000, industrial equipment: \$590,000

9. $(I - A)^{-1} \begin{bmatrix} 100 \\ 80 \\ 200 \end{bmatrix} \approx \begin{bmatrix} 398 \\ 313 \\ 452 \end{bmatrix}$

Manufacturing: \$398 million, transportation: \$313 million, agricultural: \$452 million

11. ```
round((identity(
3)-[A])⁻¹[D],2)
     [[10.25]
      [13.82]
      [8.65]]
```

## Chapter 2 Supplementary Exercises

**1.** $\begin{bmatrix} 3 & -6 & 1 \\ 2 & 4 & 6 \end{bmatrix} \xrightarrow{\frac{1}{3}[1]} \begin{bmatrix} 1 & -2 & \frac{1}{3} \\ 2 & 4 & 6 \end{bmatrix}$

$\xrightarrow{[2]+(-2)[1]} \begin{bmatrix} 1 & -2 & \frac{1}{3} \\ 0 & 8 & \frac{16}{3} \end{bmatrix}$

**2.** $\begin{bmatrix} -5 & -3 & 1 \\ 4 & 2 & 0 \\ 0 & 6 & 7 \end{bmatrix} \xrightarrow{\frac{1}{2}[2]} \begin{bmatrix} -5 & -3 & 1 \\ 2 & 1 & 0 \\ 0 & 6 & 7 \end{bmatrix}$

$\xrightarrow{[1]+3[2]} \begin{bmatrix} 1 & 0 & 1 \\ 2 & 1 & 0 \\ 0 & 6 & 7 \end{bmatrix}$

$\xrightarrow{[3]+(-6)[2]} \begin{bmatrix} 1 & 0 & 1 \\ 2 & 1 & 0 \\ -12 & 0 & 7 \end{bmatrix}$
```

3.
$$\begin{bmatrix} \frac{1}{2} & -1 & \bigm| & -3 \\ 4 & -5 & \bigm| & -9 \end{bmatrix}$$
$$\begin{bmatrix} 1 & -2 & \bigm| & -6 \\ 0 & \underline{3} & \bigm| & 15 \end{bmatrix}$$
$$\begin{bmatrix} 1 & 0 & \bigm| & 4 \\ 0 & 1 & \bigm| & 5 \end{bmatrix}$$
$x = 4, y = 5$

4.
$$\begin{bmatrix} \underline{3} & 0 & 9 & \bigm| & 42 \\ 2 & 1 & 6 & \bigm| & 30 \\ -1 & 3 & -2 & \bigm| & -20 \end{bmatrix}$$
$$\begin{bmatrix} 1 & 0 & 3 & \bigm| & 14 \\ 0 & \underline{1} & 0 & \bigm| & 2 \\ 0 & 3 & 1 & \bigm| & -6 \end{bmatrix}$$
$$\begin{bmatrix} 1 & 0 & 3 & \bigm| & 14 \\ 0 & 1 & 0 & \bigm| & 2 \\ 0 & 0 & \underline{1} & \bigm| & -12 \end{bmatrix}$$
$$\begin{bmatrix} 1 & 0 & 0 & \bigm| & 50 \\ 0 & 1 & 0 & \bigm| & 2 \\ 0 & 0 & 1 & \bigm| & -12 \end{bmatrix}$$
$x = 50, y = 2, z = -12$

5.
$$\begin{bmatrix} 3 & -6 & 6 & \bigm| & -5 \\ -2 & 3 & -5 & \bigm| & \frac{7}{3} \\ 1 & 1 & 10 & \bigm| & 3 \end{bmatrix}$$
$$\begin{bmatrix} 1 & -2 & 2 & \bigm| & -\frac{5}{3} \\ 0 & -1 & -1 & \bigm| & -1 \\ 0 & 3 & 8 & \bigm| & \frac{14}{3} \end{bmatrix}$$
$$\begin{bmatrix} 1 & 0 & 4 & \bigm| & \frac{1}{3} \\ 0 & 1 & 1 & \bigm| & 1 \\ 0 & 0 & 5 & \bigm| & \frac{5}{3} \end{bmatrix}$$
$$\begin{bmatrix} 1 & 0 & 0 & \bigm| & -1 \\ 0 & 1 & 0 & \bigm| & \frac{2}{3} \\ 0 & 0 & 1 & \bigm| & \frac{1}{3} \end{bmatrix}$$
$x = -1, \ y = \dfrac{2}{3}, \ z = \dfrac{1}{3}$

6.
$$\begin{bmatrix} \underline{3} & 6 & -9 & \bigm| & 1 \\ 2 & 4 & -6 & \bigm| & 1 \\ 3 & 4 & 5 & \bigm| & 0 \end{bmatrix}$$
$$\begin{bmatrix} 1 & 2 & -3 & \bigm| & \frac{1}{3} \\ 0 & 0 & 0 & \bigm| & \frac{1}{3} \\ 0 & -2 & 14 & \bigm| & -1 \end{bmatrix}$$
$$\begin{bmatrix} 1 & 2 & -3 & \bigm| & \frac{1}{3} \\ 0 & \underline{-2} & 14 & \bigm| & -1 \\ 0 & 0 & 0 & \bigm| & \frac{1}{3} \end{bmatrix}$$
$$\begin{bmatrix} 1 & 0 & 11 & \bigm| & -\frac{2}{3} \\ 0 & 1 & -7 & \bigm| & \frac{1}{2} \\ 0 & 0 & 0 & \bigm| & \frac{1}{3} \end{bmatrix}$$
No solution

7.
$$\begin{bmatrix} \underline{1} & 2 & -5 & 3 & \bigm| & 16 \\ -5 & -7 & 13 & -9 & \bigm| & -50 \\ -1 & 1 & -7 & 2 & \bigm| & 9 \\ 3 & 4 & -7 & 6 & \bigm| & 33 \end{bmatrix}$$
$$\begin{bmatrix} 1 & 2 & -5 & 3 & \bigm| & 16 \\ 0 & \underline{3} & -12 & 6 & \bigm| & 30 \\ 0 & 3 & -12 & 5 & \bigm| & 25 \\ 0 & -2 & 8 & -3 & \bigm| & -15 \end{bmatrix}$$
$$\begin{bmatrix} 1 & 0 & 3 & -1 & \bigm| & -4 \\ 0 & 1 & -4 & 2 & \bigm| & 10 \\ 0 & 0 & 0 & \underline{-1} & \bigm| & -5 \\ 0 & 0 & 0 & 1 & \bigm| & 5 \end{bmatrix}$$
$$\begin{bmatrix} 1 & 0 & 3 & 0 & \bigm| & 1 \\ 0 & 1 & -4 & 0 & \bigm| & 0 \\ 0 & 0 & 0 & 1 & \bigm| & 5 \\ 0 & 0 & 0 & 0 & \bigm| & 0 \end{bmatrix}$$
$$\begin{cases} x + 3z = 1 \\ y - 4z = 0 \\ \quad\ w = 5 \\ \quad\ 0 = 0 \end{cases}$$
$z =$ any value, $x = -3z + 1, \ y = 4z, \ w = 5$

8. $\begin{bmatrix} 5 & -10 & | & 5 \\ 3 & -8 & | & -3 \\ -3 & 7 & | & 0 \end{bmatrix}$

$\begin{bmatrix} 1 & -2 & | & 1 \\ 0 & -2 & | & -6 \\ 0 & 1 & | & 3 \end{bmatrix}$

$\begin{bmatrix} 1 & 0 & | & 7 \\ 0 & 1 & | & 3 \\ 0 & 0 & | & 0 \end{bmatrix}$

$x = 7, y = 3$

9. $\begin{bmatrix} 2+3 \\ -1+4 \\ 0+7 \end{bmatrix} = \begin{bmatrix} 5 \\ 3 \\ 7 \end{bmatrix}$

10. $\begin{bmatrix} 1\cdot3+3\cdot1+(-2)0 & 1\cdot5+3\cdot0+(-2)(-6) \\ 4\cdot3+0\cdot1+(-1)0 & 4\cdot5+0\cdot0+(-1)(-6) \end{bmatrix} = \begin{bmatrix} 6 & 17 \\ 12 & 26 \end{bmatrix}$

11. $\begin{bmatrix} 3 & 2 \\ 5 & 4 \end{bmatrix}^{-1} = \begin{bmatrix} \frac{4}{2} & -\frac{2}{2} \\ -\frac{5}{2} & \frac{3}{2} \end{bmatrix} = \begin{bmatrix} 2 & -1 \\ -\frac{5}{2} & \frac{3}{2} \end{bmatrix}$

$\begin{bmatrix} 2 & -1 \\ -\frac{5}{2} & \frac{3}{2} \end{bmatrix}\begin{bmatrix} 0 \\ 2 \end{bmatrix} = \begin{bmatrix} -2 \\ 3 \end{bmatrix}$

$x = -2, y = 3$

12. **a.** $\begin{bmatrix} 4 & -2 & 3 \\ 8 & -3 & 5 \\ 7 & -2 & 4 \end{bmatrix}\begin{bmatrix} 1 \\ 0 \\ 3 \end{bmatrix} = \begin{bmatrix} 13 \\ 23 \\ 19 \end{bmatrix}$

$x = 13, y = 23, z = 19$

b. $\begin{bmatrix} -2 & 2 & -1 \\ 3 & -5 & 4 \\ 5 & -6 & 4 \end{bmatrix}\begin{bmatrix} 0 \\ -1 \\ 2 \end{bmatrix} = \begin{bmatrix} -4 \\ 13 \\ 14 \end{bmatrix}$

$x = -4, y = 13, z = 14$

13. $\begin{bmatrix} 2 & 6 & | & 1 & 0 \\ 1 & 2 & | & 0 & 1 \end{bmatrix}$

$\begin{bmatrix} 1 & 3 & | & \frac{1}{2} & 0 \\ 0 & \underline{-1} & | & -\frac{1}{2} & 1 \end{bmatrix}$

$\begin{bmatrix} 1 & 0 & | & -1 & 3 \\ 0 & 1 & | & \frac{1}{2} & -1 \end{bmatrix}$

$\begin{bmatrix} -1 & 3 \\ \frac{1}{2} & -1 \end{bmatrix}$

14. $\begin{bmatrix} \underline{1} & 1 & 1 & | & 1 & 0 & 0 \\ 3 & 4 & 3 & | & 0 & 1 & 0 \\ 1 & 1 & 2 & | & 0 & 0 & 1 \end{bmatrix}$

$\begin{bmatrix} 1 & 1 & 1 & | & 1 & 0 & 0 \\ 0 & \underline{1} & 0 & | & -3 & 1 & 0 \\ 0 & 0 & 1 & | & -1 & 0 & 1 \end{bmatrix}$

$\begin{bmatrix} 1 & 0 & 1 & | & 4 & -1 & 0 \\ 0 & 1 & 0 & | & -3 & 1 & 0 \\ 0 & 0 & \underline{1} & | & -1 & 0 & 1 \end{bmatrix}$

$$\left[\begin{array}{ccc|ccc} 1 & 0 & 0 & 5 & -1 & -1 \\ 0 & 1 & 0 & -3 & 1 & 0 \\ 0 & 0 & 1 & -1 & 0 & 1 \end{array}\right]$$

$$\left[\begin{array}{ccc} 5 & -1 & -1 \\ -3 & 1 & 0 \\ -1 & 0 & 1 \end{array}\right]$$

15. $\begin{cases} c + w + s = 1000 \\ 28c + 40w + 32s = 30,000 \\ c - w - s = 0 \end{cases}$

$$\left[\begin{array}{ccc} 1 & 1 & 1 \\ 28 & 40 & 32 \\ 1 & -1 & -1 \end{array}\right]^{-1} \left[\begin{array}{c} 1000 \\ 30,000 \\ 0 \end{array}\right] = \left[\begin{array}{c} 500 \\ 0 \\ 500 \end{array}\right]$$

Corn: 500 acres, wheat: 0 acres, soybeans: 500 acres

16. $x = A$'s current stockpile,
$y = B$'s current stockpile,
$a = A$'s next-year's stockpile,
$b = B$'s next-year's stock pile
$.8x + .2y = a$
$.1x + .9y = b$

a. Next year: $\left[\begin{array}{cc} .8 & .2 \\ .1 & .9 \end{array}\right]\left[\begin{array}{c} 10,000 \\ 7000 \end{array}\right] = \left[\begin{array}{c} 9400 \\ 7300 \end{array}\right]$

Two years:
$$\left[\begin{array}{cc} .8 & .2 \\ .1 & .9 \end{array}\right]^{2}\left[\begin{array}{c} 10,000 \\ 7000 \end{array}\right] = \left[\begin{array}{c} 8980 \\ 7510 \end{array}\right]$$
A: 9400, 8980; B: 7300, 7510

b. Previous year:
$$\left[\begin{array}{cc} .8 & .2 \\ .1 & .9 \end{array}\right]^{-1}\left[\begin{array}{c} 10,000 \\ 7000 \end{array}\right] \approx \left[\begin{array}{c} 10,857 \\ 6571 \end{array}\right]$$
Two years ago:
$$\left(\left[\begin{array}{cc} .8 & .2 \\ .1 & .9 \end{array}\right]^{-1}\right)^{2}\left[\begin{array}{c} 10,000 \\ 7000 \end{array}\right] \approx \left[\begin{array}{c} 12,082 \\ 5959 \end{array}\right]$$
A: 10,857, 12,082; B: 6571, 5959

c. $a - b = (.8x + .2y) - (.1x + .9y)$
$= .7x - .7y = .7(x - y)$
Thus the "missile gap" of the following year is 70% the previous, so the "missile gap" decreases 30% each year.
$a + b = .9x + 1.1y < x + y$
when $.1y < .1x$ or $y < x$.
$a + b = .9x + 1.1y > x + y$ when $y > x$.

17. $18 + .1d = 25 + .05d$
$.05d = 7$
$d = 140$
(d)

18. $\begin{cases} .05j + .10k + .25l = 3.40 \\ j + k + l = 32 \\ k - 2l = 0 \end{cases}$

$$\left[\begin{array}{ccc} .05 & .1 & .25 \\ 1 & 1 & 1 \\ 0 & 1 & -2 \end{array}\right]^{-1}\left[\begin{array}{c} 3.4 \\ 32 \\ 0 \end{array}\right] = \left[\begin{array}{c} 14 \\ 12 \\ 6 \end{array}\right]$$
There are 14 J pencils.
(c)

19. $\left(\left[\begin{array}{cc} 1 & 0 \\ 0 & 1 \end{array}\right] - \left[\begin{array}{cc} .4 & .2 \\ .1 & .3 \end{array}\right]\right)^{-1}\left[\begin{array}{c} 8 \\ 12 \end{array}\right] = \left[\begin{array}{c} 20 \\ 20 \end{array}\right]$
Industry I: 20; industry II: 20

Chapter 3

Exercises 1

1. $(8, 7)$

$$\begin{cases} 6(8) + 3(7) \leq 96 \\ 8 + 7 \leq 18 \\ 2(8) + 6(7) \leq 72 \\ 8 \geq 0,\ 7 \geq 0 \end{cases}$$

$$\begin{cases} 69 \leq 96 & \text{true} \\ 15 \leq 18 & \text{true} \\ 58 \leq 72 & \text{true} \\ 8 \geq 0,\ 7 \geq 0 & \text{true} \end{cases}$$

Yes

3. $(9, 10)$

$$\begin{cases} 6(9) + 3(10) \leq 96 \\ 9 + 10 \leq 18 \\ 2(9) + 6(10) \leq 72 \\ 9 \geq 0,\ 3 \geq 0 \end{cases}$$

$$\begin{cases} 84 \leq 96 & \text{true} \\ 19 \leq 18 & \text{false} \\ 78 \leq 72 & \text{false} \\ 9 \geq 0,\ 3 \geq 0 & \text{true} \end{cases}$$

No

5. **a.**

	A	B	*Truck capacity*
Volume	4 cubic feet	3 cubic feet	300 cubic feet
Weight	100 pounds	200 pounds	10,000 pounds
Earnings	$13	$9	

b. Volume: $4x + 3y \leq 300$
Weight: $100x + 200y \leq 10{,}000$

c. $y \leq 2x$, $x \geq 0$, $y \geq 0$

d. $13x + 9y$

e. In standard form, the inequalities from (b) and (c) are:

$$\begin{cases} y \le -\dfrac{4}{3}x + 100 \\ y \le -\dfrac{1}{2}x + 50 \\ y \le 2x \\ x \ge 0,\ y \ge 0 \end{cases}$$

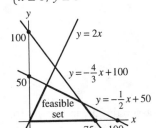

7. a.

	Essay questions	*Short-answer Questions*	*Available*
Time to answer	10 minutes	2 minutes	90 minutes
Quantity	10	50	
Required	3	10	
Worth	20 points	5 points	

b. $10x + 2y \le 90$

c. $3 \le x \le 10,\ 10 \le y \le 50$

d. $20x + 5y$

e. In standard form, the inequality from (b) is $y \le -5x + 45$. Graph the system:

$$\begin{cases} y \le -5x + 45 \\ 3 \le x \le 10 \\ 10 \le y \le 50 \end{cases}$$

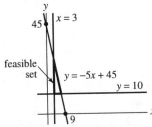

Note that the conditions $x \le 10$ and $y \le 50$ are superfluous because they are automatically assured if the other inequalities hold.

9. a.

	Alfalfa	*Corn*	*Requirements*
Protein	.13 pound	.065 pound	4550 pounds
TDN	.48 pound	.96 pound	26,880 pounds
Vitamin A	2.16 IUs	0	43,200 IUs
Cost/lb	$0.01	$0.016	

b. Protein: $.13x + .065y \geq 4550$
TDN: $.48x + .96y \geq 26,880$
Vitamin A: $2.16x \geq 43,200$
Other: $y \geq 0$
(The condition $x \geq 0$ is unnecessary because it is automatically assured if the inequality for Vitamin A holds.)

c. In standard form, the inequalities are:
$$\begin{cases} y \geq -2x + 70,000 \\ y \geq -.5x + 28,000 \\ x \geq 20,000 \\ y \geq 0 \end{cases}$$

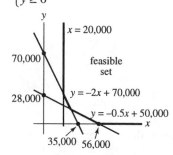

d. $.01x + .016y$

Exercises 2

1.

Vertex	$4x + 3y$
$(0, 0)$	$4(0) + 3(0) = 0$
$(0, 20)$	$4(0) + 3(20) = 60$
$(20, 0)$	$4(20) + 3(0) = 80$

The objective function is maximized at $(20, 0)$.

3. Find the vertices.

Lower left corner: $(0, 0)$

y-intercept of $y = -\dfrac{1}{2}x + 4$: $(0, 4)$

Intersection of $y = -\dfrac{1}{2}x + 4$ and $y = -x + 6$:

$$-\frac{1}{2}x + 4 = -x + 6$$

$$\frac{1}{2}x = 2$$

$$x = 4$$

$$y = -\frac{1}{2}x + 4 = -\frac{1}{2}(4) + 4 = 2$$

$(4, 2)$

x-intercept of $y = -x + 6$: $(6, 0)$

Vertex	$4x + 3y$
$(0, 0)$	$4(0) + 3(0) = 0$
$(0, 4)$	$4(0) + 3(4) = 12$
$(4, 2)$	$4(4) + 3(2) = 22$
$(6, 0)$	$4(6) + 3(0) = 24$

The objective function is maximized at $(6, 0)$.

5.

Vertex	$x + 2y$
$(0, 0)$	$0 + 2(0) = 0$
$(0, 5)$	$0 + 2(5) = 10$
$(3, 3)$	$3 + 2(3) = 9$
$(4, 0)$	$4 + 2(0) = 4$

The objective function is maximized at $(0, 5)$.

7.

Vertex	$2x + y$
$(0, 0)$	$2(0) + 0 = 0$
$(0, 5)$	$2(0) + 5 = 5$
$(3, 3)$	$2(3) + 3 = 9$
$(4, 0)$	$2(4) + 0 = 8$

The objective function is maximized at $(3, 3)$.

9.

Vertex	$8x + y$
$(0, 7)$	$8(0) + 7 = 7$
$(1, 2)$	$8(1) + 2 = 10$
$(2, 1)$	$8(2) + 1 = 17$
$(6, 0)$	$8(6) + 0 = 48$

The objective function is minimized at $(0, 7)$.

11.

Vertex	$2x + 3y$
$(0, 7)$	$2(0) + 3(7) = 21$
$(1, 2)$	$2(1) + 3(2) = 8$
$(2, 1)$	$2(2) + 3(1) = 7$
$(6, 0)$	$2(6) + 3(0) = 12$

The objective function is minimized at $(2, 1)$.

13. Find the vertices.

Lower left corner: $(0, 0)$

Intersection of $y = 2x$ and $y = -\dfrac{1}{2}x + 50$:

$$2x = -\frac{1}{2}x + 50$$

$$\frac{5}{2}x = 50$$

$$x = 20$$

$$y = 2x = 2(20) = 40$$

$(20, 40)$

Intersection of $y = -\dfrac{1}{2}x + 50$ and

$y = -\dfrac{4}{3}x + 100$:

$$-\frac{1}{2}x + 50 = -\frac{4}{3}x + 100$$

$$\frac{5}{6}x = 50$$

$$x = 60$$

$$y = -\frac{1}{2}x + 50 = -\frac{1}{2}(60) + 50 = 20$$

$(60, 20)$

x-intercept of $y = -\dfrac{4}{3}x + 100$: $(75, 0)$

Vertex	Earnings = $13x + 9y$
(0, 0)	$13(0) + 9(0) = 0$
(20, 40)	$13(20) + 9(40) = 620$
(60, 20)	$13(60) + 9(20) = 960$
(75, 0)	$13(75) + 9(0) = 975$

The earnings are maximized at (75, 0).
Ship 75 crates of cargo A and no crates of cargo B.

15.

Vertex	Score = $20x + 5y$
(3, 10)	$20(3) + 5(10) = 110$
(3, 30)	$20(3) + 5(30) = 210$
(7, 10)	$20(7) + 5(10) = 190$

The score is maximized at (3, 30).
Answer 3 essay questions and
30 short-answer questions.

17.

Vertex	Profit = $150x + 70y$
(0, 0)	$150(0) + 70(0) = 0$
(0, 12)	$150(0) + 70(12) = 840$
(9, 9)	$150(9) + 70(9) = 1980$
(14, 4)	$150(14) + 70(4) = 2380$
(16, 0)	$150(16) + 70(0) = 2400$

The profit is maximized at (16, 0).
Make 16 chairs and no sofas.

19. In standard form the inequalities are:
$$\begin{cases} y \geq -2x + 10 \\ y \geq -\dfrac{1}{2}x + 7 \\ x \geq 0, \ y \geq 0 \end{cases}$$

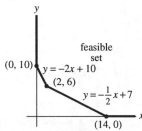

Find the vertices.
y-intercept of $y = -2x + 10$: (0, 10)

Intersection of $y = -2x + 10$ and
$$y = -\frac{1}{2}x + 7:$$
$$-\frac{1}{2}x + 7 = -2x + 10$$
$$\frac{3}{2}x = 3$$
$$x = 2$$
$$y = -2x + 10 = -2(2) + 10 = 6$$
$$(2, 6)$$

x-intercept of $y = -\dfrac{1}{2}x + 7$: (14, 0)

Vertex	$3x + 4y$
(0, 10)	$3(0) + 4(10) = 40$
(2, 6)	$3(2) + 4(6) = 30$
(14, 0)	$3(14) + 4(0) = 42$

The minimum value is 30 and occurs at (2, 6).

21. In standard form the inequalities are:
$$\begin{cases} y \leq -\dfrac{1}{2}x + 10 \\ y \geq -\dfrac{3}{2}x + 12 \\ x \leq 6 \\ x \geq 0, \ y \geq 0 \end{cases}$$

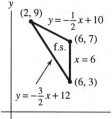

Find the vertices.

Intersection of $y = -\dfrac{1}{2}x + 10$ and
$$y = -\frac{3}{2}x + 12:$$
$$-\frac{1}{2}x + 10 = -\frac{3}{2}x + 12$$
$$x = 2$$
$$y = -\frac{1}{2}x + 10 = -\frac{1}{2}(2) + 10 = 9$$

$(2, 9)$

Intersection of $y = -\frac{1}{2}x + 10$ and $x = 6$:

$x = 6$

$y = -\frac{1}{2}x + 10 = -\frac{1}{2}(6) + 10 = 7$

$(6, 7)$

Intersection of $y = -\frac{3}{2}x + 12$ and $x = 6$:

$x = 6$

$y = -\frac{3}{2}x + 12 = -\frac{3}{2}(6) + 12 = 3$

$(6, 3)$

Vertex	$2x + 5y$
$(2, 9)$	$2(2) + 5(9) = 49$
$(6, 7)$	$2(6) + 5(7) = 47$
$(6, 3)$	$2(6) + 5(3) = 27$

The maximum value is 49 and occurs at $(2, 9)$.

23. In standard form from the inequalities are:

$$\begin{cases} y \le -\frac{1}{3}x + 40 \\ y \le -\frac{7}{2}x + 78 \\ x \le 20 \\ x \ge 0, \ y \ge 0 \end{cases}$$

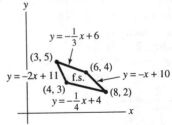

Find the vertices.

Lower left corner: $(0, 0)$

y-intercept of $y = -\frac{1}{3}x + 40$: $(0, 40)$

Intersection of $y = -\frac{1}{3}x + 40$ and

$y = -\frac{7}{2}x + 78$:

$-\frac{1}{3}x + 40 = -\frac{7}{2}x + 78$

$\frac{19}{6}x = 38$

$x = 12$

$y = -\frac{1}{3}x + 40 = -\frac{1}{3}(12) + 40 = 36$

$(12, 36)$

Intersection of $y = -\frac{7}{2}x + 78$ and $x = 20$:

$x = 20$

$y = -\frac{7}{2}x + 78 = -\frac{7}{2}(20) + 78 = 8$

$(20, 8)$

Intersection of $x = 20$ and $y = 0$: $(20, 0)$

Vertex	$100x + 150y$
$(0, 0)$	$100(0) + 150(0) = 0$
$(0, 40)$	$100(0) + 150(40) = 6000$
$(12, 36)$	$100(12) + 150(36) = 6600$
$(20, 8)$	$100(20) + 150(8) = 3200$
$(20, 0)$	$100(20) + 150(0) = 2000$

The maximum value is 6600 and occurs at $(12, 36)$.

25. The inequalities are given in standard form.

Find the vertices.

Intersection of $y = -2x + 11$ and

$y = -\frac{1}{3}x + 6$:

$-\frac{1}{3}x + 6 = -2x + 11$

$\frac{5}{3}x = 5$

$x = 3$

$$y = -\frac{1}{3}x + 6 = -\frac{1}{3}(3) + 6 = 5$$

(3, 5)

Intersection of $y = -\frac{1}{3}x + 6$ and

$y = -x + 10$:

$$-\frac{1}{3}x + 6 = -x + 10$$

$$\frac{2}{3}x = 4$$

$x = 6$

$y = -x + 10 = -6 + 10 = 4$

(6, 4)

Intersection of $y = -x + 10$ and

$y = -\frac{1}{4}x + 4$:

$$-\frac{1}{4}x + 4 = -x + 10$$

$$\frac{3}{4}x = 6$$

$x = 8$

$y = -x + 10 = -8 + 10 = 2$

(8, 2)

Intersection of $y = -\frac{1}{4}x + 4$ and

$y = -2x + 11$:

$$-\frac{1}{4}x + 4 = -2x + 11$$

$$\frac{7}{4}x = 7$$

$x = 4$

$y = -2x + 11 = -2(4) + 11 = 3$

(4, 3)

Vertex	$7x + 4y$
(3, 5)	$7(3) + 4(5) = 41$
(6, 4)	$7(6) + 4(4) = 58$
(8, 2)	$7(8) + 4(2) = 64$
(4, 3)	$7(4) + 4(3) = 40$

The minimum value is 40 and occurs at (4, 3).

27.

	Hockey games	*Soccer games*	*Available*
Assembly	2 labor-hours	3 labor-hours	42 labor-hours
Testing	2 labor-hours	1 labor-hour	26 labor-hours

Let x be the number of hockey games produced each day, and let y be the number of soccer games produced each day.

Assembly: $2x + 3y \le 42$

Testing: $2x + y \le 26$

In standard form the equations are:

$$\begin{cases} y \le -\frac{2}{3}x + 14 \\ y \le -2x + 26 \\ x \ge 0, \ y \ge 0 \end{cases}$$

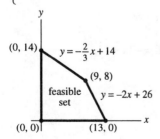

Find the vertices.
Lower left corner: $(0, 0)$

y-intercept of $y = -\dfrac{2}{3}x + 14$: $(0, 14)$

Intersection of $y = -\dfrac{2}{3}x + 14$ and $y = -2x + 26$:

$-\dfrac{2}{3}x + 14 = -2x + 26$

$\dfrac{4}{3}x = 12$

$x = 9$
$y = -2x + 26 = -2(9) + 26 = 8$
$(9, 8)$
x-intercept of $y = -2x + 26$: $(13, 0)$
The total daily output is simply the total number of games produced, or $x + y$.

Vertex	*Output* $= x + y$
$(0, 0)$	$0 + 0 = 0$
$(0, 14)$	$0 + 14 = 14$
$(9, 8)$	$9 + 8 = 17$
$(13, 0)$	$13 + 0 = 13$

The maximum output occurs at $(9, 8)$. Produce 9 hockey games and 8 soccer games each day.

29.

	First type	*Second type*	*Available*
Lots	1	1	150
Capital	$12,000	$32,000	$2,880,000
Labor	150 labor-days	200 labor-days	24,000 labor-days
Profit	$2400	$3400	

Let x be the number of homes of the first type, and let y be the number of homes of the second type.
Lots: $x + y \le 150$
Capital: $12,000x + 32,000y \le 2,880,000$
Labor: $150x + 200y \le 24,000$
In standard form the inequalities are :

$$\begin{cases} y \le -x + 150 \\ y \le -\dfrac{3}{8}x + 90 \\ y \le -\dfrac{3}{4}x + 120 \\ x \ge 0, \ y \ge 0 \end{cases}$$

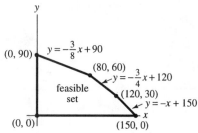

Find the vertices.

Lower left corner: $(0, 0)$

y-intercept of $y = -\dfrac{3}{8}x + 90$: $(0, 90)$

Intersection of $y = -\dfrac{3}{8}x + 90$ and $y = -\dfrac{3}{4}x + 120$:

$-\dfrac{3}{8}x + 90 = -\dfrac{3}{4}x + 120$

$\dfrac{3}{8}x = 30$

$x = 80$

$y = -\dfrac{3}{8}x + 90 = -\dfrac{3}{8}(80) + 90 = 60$

$(80, 60)$

Intersection of $y = -\dfrac{3}{4}x + 120$ and $y = -x + 150$:

$-\dfrac{3}{4}x + 120 = -x + 150$

$\dfrac{1}{4}x = 30$

$x = 120$

$y = -x + 150 = -120 + 150 = 30$

$(120, 30)$

x-intercept of $y = -x + 150$: $(150, 0)$

Vertex	Profit $= 2400x + 3400y$
$(0, 0)$	$2400(0) + 3400(0) = 0$
$(0, 90)$	$2400(0) + 3400(90) = 306{,}000$
$(80, 60)$	$2400(80) + 3400(60) = 396{,}000$
$(120, 30)$	$2400(120) + 3400(30) = 390{,}000$
$(150, 0)$	$2400(150) + 3400(0) = 360{,}000$

The maximum profit is achieved at $(80, 60)$.

Make 80 homes of the first type and 60 homes of the second type.

31.

	Fruit Delight	*Heavenly Punch*	*Available*
Pineapple juice	10 ounces	10 ounces	9000 ounces
Orange juice	3 ounces	2 ounces	2400 ounces
Apricot juice	1 ounce	2 ounces	1400 ounces
Profit	$.20	$.30	

Let x be the number of cans of Fruit Delight, and let y be the number of cans of Heavenly Punch produced each week.

Pineapple: $10x + 10y \leq 9000$

Orange: $3x + 2y \leq 2400$

Apricot: $x + 2y \leq 1400$

In standard form the inequalities are:

$$\begin{cases} y \leq -x + 900 \\ y \leq -\dfrac{3}{2}x + 1200 \\ y \leq -\dfrac{1}{2}x + 700 \\ x \geq 0,\ y \geq 0 \end{cases}$$

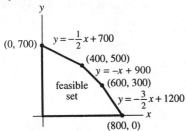

Find the vertices.

Lower left corner: $(0, 0)$

y-intercept of $y = -\dfrac{1}{2}x + 700$: $(0, 700)$

Intersection of $y = -\dfrac{1}{2}x + 700$ and $y = -x + 900$:

$$-\frac{1}{2}x + 700 = -x + 900$$

$$\frac{1}{2}x = 200$$

$$x = 400$$

$$y = -x + 900 = -400 + 900 = 500$$

$(400, 500)$

Intersection of $y = -x + 900$ and $y = -\dfrac{3}{2}x + 1200$:

$$-x + 900 = -\frac{3}{2}x + 1200$$

$\frac{1}{2}x = 300$

$x = 600$

$y = -x + 900 = -600 + 900 = 300$

$(600, 300)$

x-intercept of $y = -\frac{3}{2}x + 1200$: $(800, 0)$

Vertex	Profit = .2x + .3y
(0, 0)	.2(0) + .3(0) = 0
(0, 700)	.2(0) + .3(700) = 210
(400, 500)	.2(400) + .3(500) = 230
(600, 300)	.2(600) + .3(300) = 210
(800, 0)	.2(800) + .3(0) = 160

The maximum profit is achieved at (400, 500).
Make 400 cans of Fruit Delight and 500 cans of Heavenly Punch.

33. Since the former can spend $2400 for labor at $8 per hour, the available labor is $2400 \div 8 = 300$ hours.

	Oats	Corn	Available
Capital	$18	$36	$2100
Labor	2 hours	6 hours	300 hours
Land	1 acre	1 acre	100 acres
Revenue	$55	$125	

Let x be the number of acres of oats, and let y be the number of acres of corn.
Capital: $18x + 36y \leq 2100$
Labor: $2x + 6y \leq 300$
Land: $x + y \leq 100$
In standard form, the inequalities are:

$$\begin{cases} y \leq -\frac{1}{2}x + \frac{175}{3} \\ y \leq -\frac{1}{3}x + 50 \\ y \leq -x + 100 \\ x \geq 0, \ y \geq 0 \end{cases}$$

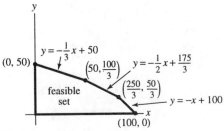

Find the vertices.

Lower left corner: $(0, 0)$

y-intercept of $y = -\frac{1}{3}x + 50$: $(0, 50)$

Intersection of $y = -\frac{1}{3}x + 50$ and $y = -\frac{1}{2}x + \frac{175}{3}$:

$$-\frac{1}{3}x + 50 = -\frac{1}{2}x + \frac{175}{3}$$

$$\frac{1}{6}x = \frac{25}{3}$$

$$x = 50$$

$$y = -\frac{1}{3}x + 50 = -\frac{1}{3}(50) + 50 = \frac{100}{3}$$

$$\left(50, \frac{100}{3}\right)$$

Intersection of $y = -\frac{1}{2}x + \frac{175}{3}$ and $y = -x + 100$:

$$-\frac{1}{2}x + \frac{175}{3} = -x + 100$$

$$\frac{1}{2}x = \frac{125}{3}$$

$$x = \frac{250}{3}$$

$$y = -x + 100 = -\frac{250}{3} + 100 = \frac{50}{3}$$

$$\left(\frac{250}{3}, \frac{50}{3}\right)$$

x-intercept of $y = -x + 100$: $(100, 0)$

Find the objective function for the profit.

Revenue: $55x + 125y$

Leftover capital: $2100 - 18x - 36y$

Leftover labor cash reserve: $2400 - 8(2x + 6y) = 2400 - 16x - 48y$

The profit is the sum of the above: $21x + 41y + 4500$

Vertex	$Profit = 21x + 41y + 4500$
$(0, 0)$	$21(0) + 41(0) + 4500 = 4500$
$(0, 50)$	$21(0) + 41(50) + 4500 = 6550$
$\left(50, \frac{100}{3}\right)$	$21(50) + 41\left(\frac{100}{3}\right) + 4500 \approx 6916.67$
$\left(\frac{250}{3}, \frac{50}{3}\right)$	$21\left(\frac{250}{3}\right) + 41\left(\frac{50}{3}\right) + 4500 \approx 6933.33$
$(100, 0)$	$21(100) + 41(0) + 4500 = 6600$

The maximum profit is achieved at $\left(\dfrac{250}{3}, \dfrac{50}{3}\right)$, or $\left(83\dfrac{1}{3}, 16\dfrac{2}{3}\right)$. The farmer should plant $83\dfrac{1}{3}$ acres of

oats and $16\dfrac{2}{3}$ acres of corn to make a profit of \$6933.33.

35. a.

	I_1	I_2	Available
M_1	3 ounces	4 ounces	40 ounces
M_2	2 ounces	1 ounce	20 ounces
M_3	2 ounces	3 ounces	60 ounces
Profit	\$8	\$6	

Let x be the number of item I_1 made each day, and let y be the number of item I_2.

M_1: $3x + 4y \le 40$

M_2: $2x + y \le 20$

M_3: $2x + 3y \le 60$

In standard form, the inequalities are:

$$\begin{cases} y \le -\dfrac{3}{4}x + 10 \\ y \le -2x + 20 \\ y \le -\dfrac{2}{3}x + 20 \\ x \ge 0,\ y \ge 0 \end{cases}$$

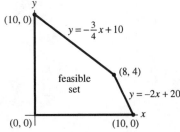

Since $x \ge 0$, the inequality $y \le -2x + 20$ assures that $y \le -\dfrac{2}{3}x + 20$ is satisfied. Therefore, the inequality $y \le -\frac{2}{3}x + 20$ is superfluous.

Find the vertices:
Lower left corner: (0, 0)

y-intercept of $y = -\frac{3}{4}x + 10$: (0, 10)

Intersection of $y = -\frac{3}{4}x + 10$ and $y = -2x + 20$:

$-\frac{3}{4}x + 10 = -2x + 20$

$\frac{5}{4}x = 10$

$x = 8$
$y = -2x + 20 = -2(8) + 20 = 4$
(8, 4)
x-intercept of $y = -2x + 20$: (10, 0)

Vertex	Profit = 8x + 6y
(0, 0)	8(0) + 6(0) = 0
(0, 10)	8(0) + 6(10) = 60
(8, 4)	8(8) + 6(4) = 88
(10, 0)	8(10) + 6(0) = 80

The maximum profit is achieved at (8, 4).
Make 8 of item I_1 and 4 of item I_2.

b. $88

c. M_1: $3x + 4y = 3(8) + 4(4) = 40$
M_2: $2x + y = 2(8) + 4 = 20$
M_3: $2x + 3y = 2(8) + 3(4) = 28$
40 ounces of M_1, 20 ounces of M_2, and 28 ounces of M_3 are used.

d.

Vertex	Profit = 13x + 6y
(0, 0)	13(0) + 6(0) = 0
(0, 10)	13(0) + 6(10) = 60
(8, 4)	13(8) + 6(4) = 128
(10, 0)	13(10) + 6(0) = 130

The maximum profit is achieved at (10, 0).
Make 10 of item I_1 and 0 of item I_2.

37. Since $x \ge 0$ and $y \ge 0$, $4x + 3y \ge x + y$. Therefore, the inequality $x + y \ge 6$ implies that $4x + 3y \ge 6$, which contradicts $4x + 3y \le 4$.
The feasible set contains no points.

Exercises 3

1. a. $21x + 14y = c$

$$y = -\frac{3}{2}x + \frac{c}{14}$$

b. Up

c. B

3. The objective function $ax + by$ has constant value on any line of slope $-\frac{a}{b}$. The slope of the line containing $(8, 3)$ and $(9, 0)$ is $\frac{0-3}{9-8} = -3$. Therefore, if $-\frac{a}{b} < 0$, we require $-\frac{a}{b} \le -3$, or $\frac{a}{b} \ge 3$, where $a > 0$ and $b > 0$. One possibility is $a = 5$ and $b = 1$, giving the objective function $5x + y$. (We could also have chosen an objective function that is constant on lines of nonnegative or undefined slope of the form $ax + by$ with $a \ge 0$ and $b \le 0$.)

5. The objective function $ax + by$ has constant value on any line of slope $-\frac{a}{b}$. Since the slope of the line containing $(3, 8)$ and $(8, 3)$ is $\frac{3-8}{8-3} = -1$, and the slope of the line containing $(8, 3)$ and $(9, 0)$ is $\frac{0-3}{9-8} = -3$, we require $-1 \ge -\frac{a}{b} \ge -3$, or $1 \le \frac{a}{b} \le 3$. One possibility is $a = 2$ and $b = 1$, giving the objective function $2x + y$.

7. The objective function $ax + by$ has constant value on any line of slope $-\frac{a}{b}$. The slope of the line containing $(6, 1)$ and $(9, 0)$ is $\frac{0-1}{9-6} = -\frac{1}{3}$.

We require $0 \ge -\frac{a}{b} \ge -\frac{1}{3}$, or $0 \le \frac{a}{b} \le \frac{1}{3}$, where $a \ge 0$ and $b > 0$. One possibility is

$a = 1$ and $b = 5$, giving the objective function $x + 5y$.

9. The objective function $ax + by$ has constant value on any line of slope $-\frac{a}{b}$. The slope of the line containing $(1, 6)$ and $(6, 1)$ is $\frac{1-6}{6-1} = -1$, and the slope of the line containing $(6, 1)$ and $(9, 0)$ is $\frac{0-1}{9-6} = -\frac{1}{3}$.

We require $-\frac{1}{3} \ge -\frac{a}{b} \ge -1$, or $\frac{1}{3} \le \frac{a}{b} \le 1$, where $a > 0$ and $b > 0$. One possibility is $a = 2$ and $b = 3$, giving the objective function $2x + 3y$.

11. The objective function $3x + 2y$ has constant value on any line of slope $-\frac{3}{2}$. Since this is between -1 and -4, the objective function is maximized at C.

13. The objective function $10x + 2y$ has constant value on any line of slope -5. Since this is less (steeper) than -4, the objective function is maximized at D.

15. The objective function $2x + 10y$ has constant value on any line of slope $-\frac{1}{5}$. Since this is between 0 and $-\frac{1}{4}$, the objective function is minimized at D.

17. The objective function $2x + 3y$ has constant value on any line of slope $-\frac{2}{3}$. Since this is between $-\frac{1}{4}$ and -1, the objective function is minimized at C.

19. The objective function $x + ky$ has constant value on any line of slope $-\dfrac{1}{k}$. The slope of the line

containing $(0, 5)$ and $(3, 4)$ is $\dfrac{4-5}{3-0} = -\dfrac{1}{3}$, and the slope of the line containing $(3, 4)$ and $(4, 0)$ is

$\dfrac{0-4}{4-3} = -4$. We require $-4 \le -\dfrac{1}{k} \le -\dfrac{1}{3}$. This is equivalent to $4 \ge \dfrac{1}{k} \ge \dfrac{1}{3}$, or $\dfrac{1}{4} \le k \le 3$.

21.

	Brand A	*Brand B*	*Requirement*
Protein	3 units	1 unit	6 units
Carbohydrate	1 unit	1 unit	4 units
Fat	2 units	6 units	12 units
Cost	$.80	$.50	

Let x be the number of units of brand A, and let y be the number of units of brand B.
Protein: $3x + y \ge 6$
Carbohydrate: $x + y \ge 4$
Fat: $2x + 6y \ge 12$
In standard form the inequalities are:
$$\begin{cases} y \ge -3x + 6 \\ y \ge -x + 4 \\ y \ge -\dfrac{1}{3}x + 2 \\ x \ge 0,\ y \ge 0 \end{cases}$$

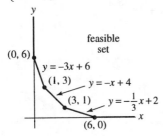

Vertex	*Cost* = $.8x + .5y$
$(0, 6)$	3
$(1, 3)$	2.3
$(3, 1)$	2.9
$(6, 0)$	4.8

The minimum cost of $2.30 is obtained at $(1, 3)$.
Feed 1 can of brand A and 3 cans of brand B.

23.

	Oranges	Grapefruits	Avocados
Profit	$5	$6	$4
Variables	x	y	$100 - x - y$

The required inequalities are:

$$\begin{cases} x \geq 20 \\ y \geq 10 \\ 100 - x - y \geq 30 \\ x \geq y \\ x \geq 0,\ y \geq 0 \\ 100 - x - y \geq 0 \end{cases} \quad \text{or} \quad \begin{cases} x \geq 20 \\ y \geq 10 \\ y \leq -x + 70 \\ y \leq x \\ x \geq 0,\ y \geq 0 \\ y \leq -x + 100 \end{cases}$$

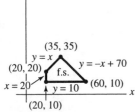

(The inequalities $x \geq 0$, $y \geq 0$, and $y \leq -x + 100$ do not appear in the graph, as they are assured by the other inequalities.)

The objective function for profit is $5x + 6y + 4(100 - x - y)$, or $400 + x + 2y$.

Vertex	Profit $= 400 + x + 2y$
(20, 10)	440
(20, 20)	460
(35, 35)	505
(60, 10)	480

The maximum profit of $505 is achieved at (35, 35). Then $100 - x - y = 100 - 35 - 35 = 30$.
Ship 35 crates of oranges, 35 crates of grapefruits, and 30 crates of avocados.

25.

	Detroit cars	Detroit trucks	Cleveland cars	Cleveland trucks
Cost	$1200	$2100	$1000	$2000
Variables	x	y	$600 - x$	$300 - y$

Since the order is a rush order, we set up constraints according to the maximum daily production of each plant. The required inequalities are:

$$\begin{cases} x + y \le 800 \\ (600 - x) + (300 - y) \le 500 \\ \qquad\qquad x \ge 0,\ y \ge 0 \\ 600 - x \ge 0 \\ 300 - y \ge 0 \end{cases} \text{ or } \begin{cases} y \le -x + 800 \\ y \ge -x + 400 \\ x \ge 0,\ y \ge 0 \\ x \le 600 \\ y \le 300 \end{cases}$$

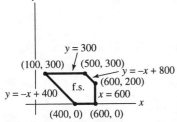

(The inequality $x \ge 0$ does not appear in the graph, as it is assured by the other inequalities.)
The objective function for the cost is $1200x + 2100y + 1000(600 - x) + 2000(300 - y)$, or
$1,200,000 + 200x + 100y$.

Vertex	Cost = 1,200,000 + 200x + 100y
(100, 300)	1,250,000
(500, 300)	1,330,000
(600, 200)	1,340,000
(600, 0)	1,320,000
(400, 0)	1,280,000

The minimum cost of $1,250,000 is achieved at (100, 300).
Then $600 - x = 600 - 100 = 500$ and $300 - y = 300 - 300 = 0$.
In Detroit make 100 cars and 300 trucks. In Cleveland make 500 cars and 0 trucks.

27. Let the variables represent the number of thousands of gallons.

	Gasoline	Jet fuel	Diesel fuel
Profit	$.15	$.12	$.10
Variables	x	y	$100 - x - y$

The required inequalities are:

$$\begin{cases} x \ge 5,\ y \ge 5 \\ 100 - x - y \ge 5 \\ x + y \ge 20 \\ x + (100 - x - y) \ge 50 \end{cases} \text{ or } \begin{cases} x \ge 5,\ y \ge 5 \\ y \le -x + 95 \\ y \ge -x + 20 \\ y \le 50 \end{cases}$$

(Note that the inequalities above have ignored the somewhat subtle issue of whether there will be enough gasoline for *both* the airline and the trucking firm at the same time. This is only a potential issue if both suppliers require gasoline—that is, if $y \le 20$ and $100 - x - y \le 50$. In this case, we require that the gasoline requirements of each firm are less than the amount of gasoline actually produced—that is,

$(20 - y) + (50 - (100 - x - y)) \le x$. This inequality is equivalent to $-30 \le 0$, so it is always satisfied.)

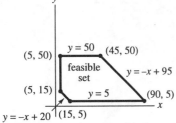

The objective function for the profit is $.15x + .12y + .1(100 - x - y)$, or $10 + .05x + .02y$.

Vertex	Profit $= 10 + .05x + .02y$
(5, 50)	11.25
(45, 50)	13.25
(90, 5)	14.6
(15, 5)	10.85
(5, 15)	10.55

The maximum profit of \$14,600 is achieved at (90, 5). Then $100 - x - y = 100 - 90 - 5 = 5$.
Produce 90,000 gallons of gasoline, 5000 gallons of jet fuel, and 5000 gallons of diesel fuel.

29.

	High-capacity	Low-capacity	Available
Cost (\$thousands)	50	30	1080
Drivers	1	1	30
Capacity	320 cases	200 cases	

Let x be the number of high-capacity trucks, and let y be the number of low-capacity trucks. The required inequalities are:

$$\begin{cases} 50x + 30y \le 1080 \\ x + y \le 30 \\ x \le 15 \\ x \ge 0, \ y \ge 0 \end{cases} \quad \text{or} \quad \begin{cases} y \le \dfrac{5}{3}x + 36 \\ y \le -x + 30 \\ x \le 15 \\ x \ge 0, \ y \ge 0 \end{cases}$$

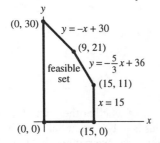

Vertex	$Capacity = 320x + 200y$
$(0, 0)$	0
$(0, 30)$	6000
$(9, 21)$	7080
$(15, 11)$	7000
$(15, 0)$	4800

The maximum capacity of 7080 cases is achieved at $(9, 21)$.
Buy 9 high-capacity trucks and 21 low-capacity trucks.

Chapter 3 Supplementary Exercises

1.

	Type A	Type B	Required or available
Passengers	50	300	1400
Flight attendants	3	4	42
Cost	$14,000	$90,000	

Let x be the number of type A planes and let y be the number of type B planes.
The required inequalities are:

$$\begin{cases} 50x + 300y \ge 1400 \\ 3x + 4y \le 42 \\ x \ge y \\ x \ge 0,\ y \ge 0 \end{cases} \quad \text{or} \quad \begin{cases} y \ge -\dfrac{1}{6}x + \dfrac{14}{3} \\ y \le -\dfrac{3}{4}x + \dfrac{21}{2} \\ y \le x \\ x \ge 0,\ y \ge 0 \end{cases}$$

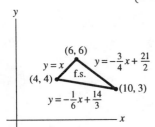

(The inequalities $x \ge 0$ and $y \ge 0$ are not shown in the graph because they are assured by the other inequalities.)

Vertex	$Cost = 14,000x + 90,000y$
$(4, 4)$	416,000
$(6, 6)$	624,000
$(10, 3)$	410,000

The minimum cost of $410,000 is achieved at $(10, 3)$.
Use 10 type A planes and 3 type B planes.

2.

	Wheat germ	*Enriched oat flour*	
Niacin	2 milligrams	3 milligrams	7 milligrams
Iron	3 milligrams	3 milligrams	9 milligrams
Thiamin	.5 milligram	.25 milligrams	1 milligram
Cost	3 cents	4 cents	

Let x be the number of ounces of wheat germ, and let y be the number of ounces of enriched oat flour. The required inequalities are:

$$\begin{cases} 2x+3y \ge 7 \\ 3x+3y \ge 9 \\ .5x+.25y \ge 1 \\ x \ge 0,\ y \ge 0 \end{cases} \text{ or } \begin{cases} y \ge -\dfrac{2}{3}x+\dfrac{7}{3} \\ y \ge -x+3 \\ y \ge -2x+4 \\ x \ge 0,\ y \ge 0 \end{cases}$$

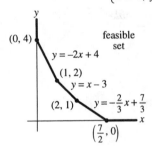

Vertex	$Cost = 3x + 4y$
$(0, 4)$	16
$(1, 2)$	11
$(2, 1)$	10
$\left(\frac{7}{2}, 0\right)$	10.5

The minimum cost of 10 cents is achieved at $(2, 1)$.
Use 2 ounces of wheat germ and 1 ounce of enriched oat flour.

3.

	Hardtops	*Sports cars*	*Available*
Assemble	8 labor-hours	18 labor-hours	360 labor-hours
Paint	2 labor-hours	2 labor-hours	50 labor-hours
Upholstery	2 labor-hours	1 labor-hour	40 labor-hours
Profit	$90	$100	

The required inequalities are:

$$\begin{cases} 8x + 18y \le 360 \\ 2x + 2y \le 50 \\ 2x + y \le 40 \\ x \ge 0,\ y \ge 0 \end{cases} \quad \text{or} \quad \begin{cases} y \le -\dfrac{4}{9}x + 20 \\ y \le -x + 25 \\ y \le -2x + 40 \\ x \ge 0,\ y \ge 0 \end{cases}$$

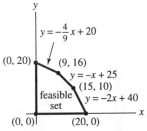

Vertex	Profit $= 90x + 100y$
(0, 0)	0
(0, 20)	2000
(9, 16)	2410
(15, 10)	2350
(20, 0)	1800

The maximum profit of $2410 is achieved at (9, 16).
Produce 9 hardtops and 16 sports cars.

4.

	Mixture A	Mixture B	Available
Peanuts	6 ounces	12 ounces	5400 ounces
Raisins	1 ounce	3 ounces	1200 ounces
Cashews	4 ounces	2 ounces	2400 ounces
Revenue	$.50	$.90	

Let x be the number of boxes of mixture A, and let y be the number of boxes of mixture B.
The required inequalities are:

$$\begin{cases} 6x + 12y \le 5400 \\ x + 3y \le 1200 \\ 4x + 2y \le 2400 \\ x \ge 0,\ y \ge 0 \end{cases} \quad \text{or} \quad \begin{cases} y \le -\dfrac{1}{2}x + 450 \\ y \le -\dfrac{1}{3}x + 400 \\ y \le -2x + 1200 \\ x \ge 0,\ y \ge 0 \end{cases}$$

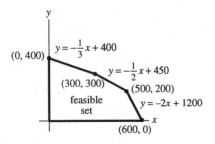

Vertex	Revenue = $.5x + .9y$
$(0, 0)$	0
$(0, 400)$	360
$(300, 300)$	420
$(500, 200)$	430
$(600, 0)$	300

The maximum revenue of $430 occurs at (500, 200).
Make 500 boxes of mixture A and 200 boxes of mixture B.

5.

	Elementary	Intermediate	Advanced
Profit	$8000	$7000	$1000
Variables	x	y	$72 - x - y$

The required inequalities are:

$$\begin{cases} 72 - x - y \geq 4 \\ x \geq 3y \\ y \geq 2(72 - x - y) \\ x \geq 0, \ y \geq 0 \end{cases} \text{ or } \begin{cases} y \leq -x + 68 \\ y \leq \dfrac{1}{3}x \\ y \leq -\dfrac{2}{3}x + 48 \\ x \geq 0, \ y \geq 0 \end{cases}$$

The objective function for the annual profit is $8000x + 7000y + 1000(72 - x - y)$, or $72{,}000 + 7000x + 6000y$.

Vertex	Profit = $72{,}000 + 7000x + 6000y$
$(48, 16)$	504,000
$(51, 17)$	531,000
$(60, 8)$	540,000

The maximum annual profit of $540,000 is achieved at (60, 8).
Then $72 - x - y = 72 - 60 - 8 = 4$.
Publish 60 elementary books, 8 intermediate books, and 4 advanced books.

6.

	Rochester	Queens	Available
Transport time	15 hours	20 hours	2100
Cost	$15	$30	$3000
Profit	$40	$30	

Let x be the number of computers sent from Rochester, and let y be the number of computers sent from Queens. The required inequalities are:

$$\begin{cases} 15x + 20y \le 2100 \\ 15x + 30y \le 3000 \\ x \le 80 \\ y \le 120 \\ x \ge 0, \ y \ge 0 \end{cases} \text{ or } \begin{cases} y \le -\dfrac{3}{4}x + 105 \\ y \le -\dfrac{1}{2}x + 100 \\ x \le 80 \\ y \le 120 \\ x \ge 0, \ y \ge 0 \end{cases}$$

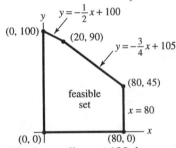

(The inequality $y \le 120$ does not appear in the graph, as it is assured by the other inequalities.)

Vertex	Profit $= 40x + 30y$
(0, 0)	0
(0, 100)	3000
(20, 90)	3500
(80, 45)	4550
(80, 0)	3200

The maximum profit of $4550 is achieved at (80, 45).
Transport 80 computers from Rochester and 45 computers from Queens.

7.

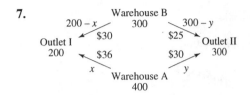

The required inequalities are:

$$\begin{cases} x \geq 0, \ y \geq 0 \\ 200 - x \geq 0 \\ 300 - y \geq 0 \\ x + y \leq 400 \\ (200 - x) + (300 - y) \leq 300 \end{cases} \quad \text{or} \quad \begin{cases} x \geq 0, \ y \geq 0 \\ x \leq 200 \\ y \leq 300 \\ y \leq -x + 400 \\ y \geq -x + 200 \end{cases}$$

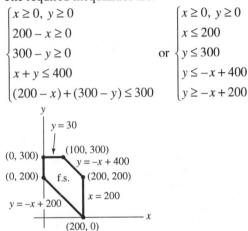

The objective function for the cost is $36x + 30y + 30(200 - x) + 25(300 - y)$, or $13,500 + 6x + 5y$.

Vertex	Cost = 13,500 + 6x + 5y
(0, 200)	14,500
(0, 300)	15,000
(100, 300)	15,600
(200, 200)	15,700
(200, 0)	14,700

The minimum cost of $15,000 is achieved at (0, 200). Then $200 - x = 200 - 0 = 200$, and $300 - y = 300 - 200 = 100$.

Transport 200 computers from warehouse A to outlet II, 200 computers from warehouse B to outlet I, and 100 computers from warehouse B to outlet II.

Chapter 4

Exercises 1

1. $\begin{cases} 20x + 30y + u = 3500 \\ 50x + 10y + v = 5000 \\ -8x - 13y + M = 0 \end{cases}$

Maximize M given $x \geq 0, y \geq 0, u \geq 0, v \geq 0$.

3. $\begin{cases} x + y + z + u = 100 \\ 3x + z + v = 200 \\ 5x + 10y + w = 100 \\ -x - 2y + 3z + M = 0 \end{cases}$

Maximize M given $x \geq 0, y \geq 0, z \geq 0, u \geq 0, v \geq 0, w \geq 0$.

5. a.
$$\begin{array}{ccccc} x & y & u & v & M \\ \end{array}$$
$$\begin{bmatrix} 20 & 30 & 1 & 0 & 0 & | & 3500 \\ 50 & 10 & 0 & 1 & 0 & | & 5000 \\ -8 & -13 & 0 & 0 & 1 & | & 0 \end{bmatrix}$$

b. $x = 0, y = 0, u = 3500, v = 5000, M = 0$

7. a.
$$\begin{array}{ccccccc} x & y & z & u & v & w & M \end{array}$$
$$\begin{bmatrix} 1 & 1 & 1 & 1 & 0 & 0 & 0 & | & 100 \\ 3 & 0 & 1 & 0 & 1 & 0 & 0 & | & 200 \\ 5 & 10 & 0 & 0 & 0 & 1 & 0 & | & 100 \\ -1 & -2 & 3 & 0 & 0 & 0 & 1 & | & 0 \end{bmatrix}$$

b. $x = 0, y = 0, z = 0, u = 100, v = 200,$
$w = 100, M = 0$

9. $x = 15, y = 0, u = 10, v = 0, M = 20$

11. $x = 10, y = 0, z = 15, u = 23, v = 0, w = 0,$
$M = -11$

13. a.
$$\begin{array}{ccccc} x & y & u & v & M \end{array}$$
$$\begin{bmatrix} 1 & \frac{3}{2} & \frac{1}{2} & 0 & 0 & | & 6 \\ 0 & -\frac{1}{2} & -\frac{1}{2} & 1 & 0 & | & 4 \\ 0 & -5 & 5 & 0 & 1 & | & 60 \end{bmatrix}$$
$x = 6, y = 0, u = 0, v = 4, M = 60$

b.
$$\begin{array}{ccccc} x & y & u & v & M \end{array}$$
$$\begin{bmatrix} \frac{2}{3} & 1 & \frac{1}{3} & 0 & 0 & | & 4 \\ \frac{1}{3} & 0 & -\frac{1}{3} & 1 & 0 & | & 6 \\ \frac{10}{3} & 0 & \frac{20}{3} & 0 & 1 & | & 80 \end{bmatrix}$$
$x = 0, y = 4, u = 0, v = 6, M = 80$

c.
$$\begin{array}{ccccc} x & y & u & v & M \end{array}$$
$$\begin{bmatrix} 0 & 1 & 1 & -2 & 0 & | & -8 \\ 1 & 1 & 0 & 1 & 0 & | & 10 \\ 0 & -10 & 0 & 10 & 1 & | & 100 \end{bmatrix}$$
$x = 10, y = 0, u = -8, v = 0, M = 100$

d.
$$\begin{array}{ccccc} x & y & u & v & M \end{array}$$
$$\begin{bmatrix} -1 & 0 & 1 & -3 & 0 & | & -18 \\ 1 & 1 & 0 & 1 & 0 & | & 10 \\ 10 & 0 & 0 & 20 & 1 & | & 200 \end{bmatrix}$$
$x = 0, y = 10, u = -18, v = 0, M = 200$

15. M becomes greatest after operation (d).

Exercises 2

1. a. -12 is the most negative entry in the last row, and $\dfrac{6}{3} < \dfrac{10}{2}$. Pivot about the 3.

b.

	x	y	u	v	M	
u	$\frac{16}{3}$	0	1	$-\frac{2}{3}$	0	6
y	$\frac{1}{3}$	1	0	$\frac{1}{3}$	0	2
M	0	0	0	4	1	24

c. $x = 0, y = 2, u = 6, v = 0, M = 24$

3. a. -2 is the only negative entry in the last row, and $\frac{5}{10} < \frac{12}{12}$. Pivot about the 10.

b.

	x	y	u	v	M	
u	-13	0	1	$-\frac{6}{5}$	0	6
y	$\frac{3}{2}$	1	0	$\frac{1}{10}$	0	$\frac{1}{2}$
M	7	0	0	$\frac{1}{5}$	1	1

c. $x = 0, \ y = \frac{1}{2}, \ u = 6, v = 0, M = 1$

In 5–25, the pivot elements are underlined.

5.

	x	y	u	v	M	
u	1	1	1	0	0	7
v	1	2	0	1	0	10
M	-1	-3	0	0	1	0

	x	y	u	v	M	
u	$\frac{1}{2}$	0	1	$-\frac{1}{2}$	0	2
y	$\frac{1}{2}$	1	0	$\frac{1}{2}$	0	5
M	$\frac{1}{2}$	0	0	$\frac{3}{2}$	1	15

$x = 0, y = 5; M = 15$

7.

	x	y	u	v	M	
u	5	1	1	0	0	80
v	3	2	0	1	0	76
M	-4	-2	0	0	1	0

	x	y	u	v	M	
x	1	$\frac{1}{5}$	$\frac{1}{5}$	0	0	16
v	0	$\frac{7}{5}$	$-\frac{3}{5}$	1	0	28
M	0	$-\frac{6}{5}$	$\frac{4}{5}$	0	1	64

	x	y	u	v	M	
x	1	0	$\frac{2}{7}$	$-\frac{1}{7}$	0	12
y	0	1	$-\frac{3}{7}$	$\frac{5}{7}$	0	20
M	0	0	$\frac{2}{7}$	$\frac{6}{7}$	1	88

$x = 12, y = 20; M = 88$

9.

	x	y	z	u	v	M	
u	1	0	2	1	0	0	10
v	0	3	1	0	1	0	24
M	-1	-3	-5	0	0	1	0

	x	y	z	u	v	M	
z	$\frac{1}{2}$	0	1	$\frac{1}{2}$	0	0	5
v	$-\frac{1}{2}$	3	0	$-\frac{1}{2}$	1	0	19
M	$\frac{3}{2}$	-3	0	$\frac{5}{2}$	0	1	25

	x	y	z	u	v	M	
z	$\frac{1}{2}$	0	1	$\frac{1}{2}$	0	0	5
y	$-\frac{1}{6}$	1	0	$-\frac{1}{6}$	$\frac{1}{3}$	0	$\frac{19}{3}$
M	1	0	0	2	1	1	44

$x = 0, \ y = \frac{19}{3}, \ z = 5; M = 44$

11.

	x	y	u	v	w	M	
u	5	1	1	0	0	0	30
v	3	2	0	1	0	0	60
w	1	1	0	0	1	0	50
M	-2	-3	0	0	0	1	0

	x	y	u	v	w	M	
y	5	1	1	0	0	0	30
v	-7	0	-2	1	0	0	0
w	-4	0	-1	0	1	0	20
M	13	0	3	0	0	1	90

$x = 0, y = 30; M = 90$
Pivoting about the 2 instead gives the same solution although the tableau is different.

13.

$$\begin{array}{c} \\ u \\ v \\ M \end{array}\begin{array}{c}\begin{array}{cccccc} x & y & u & v & M \end{array}\\ \left[\begin{array}{ccccc|c} 2 & 3 & 1 & 0 & 0 & 400 \\ 1 & 1 & 0 & 1 & 0 & 150 \\ \hline -6 & -7 & 0 & 0 & 1 & 300 \end{array}\right]\end{array}$$

$$\begin{array}{c} \\ y \\ v \\ M \end{array}\begin{array}{c}\begin{array}{cccccc} x & y & u & v & M \end{array}\\ \left[\begin{array}{ccccc|c} \frac{2}{3} & 1 & \frac{1}{3} & 0 & 0 & \frac{400}{3} \\ \frac{1}{3} & 0 & -\frac{1}{3} & 1 & 0 & \frac{50}{3} \\ \hline -\frac{4}{3} & 0 & \frac{7}{3} & 0 & 1 & \frac{3700}{3} \end{array}\right]\end{array}$$

$$\begin{array}{c} \\ y \\ x \\ M \end{array}\begin{array}{c}\begin{array}{cccccc} x & y & u & v & M \end{array}\\ \left[\begin{array}{ccccc|c} 0 & 1 & 1 & -2 & 0 & 100 \\ 1 & 0 & -1 & 3 & 0 & 50 \\ \hline 0 & 0 & 1 & 4 & 1 & 1300 \end{array}\right]\end{array}$$

$x = 50$, $y = 100$; $M = 1300$

15. Let c be the number of chairs, s be the number of sofas, and t be the number of tables manufactured each day. Maximize $80c + 70s + 120t$ subject to the constraints

$$\begin{cases} 6c + 3s + 8t \le 768 \\ c + s + 2t \le 144 \\ 2c + 5s \le 216 \\ c \ge 0,\ s \ge 0,\ t \ge 0 \end{cases}$$

$$\begin{array}{c} \\ u \\ v \\ w \\ M \end{array}\begin{array}{c}\begin{array}{ccccccc} c & s & t & u & v & w & M \end{array}\\ \left[\begin{array}{ccccccc|c} 6 & 3 & 8 & 1 & 0 & 0 & 0 & 768 \\ 1 & 1 & 2 & 0 & 1 & 0 & 0 & 144 \\ 2 & 5 & 0 & 0 & 0 & 1 & 0 & 216 \\ \hline -80 & -70 & -120 & 0 & 0 & 0 & 1 & 0 \end{array}\right]\end{array}$$

$$\begin{array}{c} \\ u \\ t \\ w \\ M \end{array}\begin{array}{c}\begin{array}{ccccccc} c & s & t & u & v & w & M \end{array}\\ \left[\begin{array}{ccccccc|c} 2 & -1 & 0 & 1 & -4 & 0 & 0 & 192 \\ \frac{1}{2} & \frac{1}{2} & 1 & 0 & \frac{1}{2} & 0 & 0 & 72 \\ 2 & 5 & 0 & 0 & 0 & 1 & 0 & 216 \\ \hline -20 & -10 & 0 & 0 & 60 & 0 & 1 & 8640 \end{array}\right]\end{array}$$

$$\begin{array}{c} \\ c \\ t \\ w \\ M \end{array}\begin{array}{c}\begin{array}{ccccccc} c & s & t & u & v & w & M \end{array}\\ \left[\begin{array}{ccccccc|c} 1 & -\frac{1}{2} & 0 & \frac{1}{2} & -2 & 0 & 0 & 96 \\ 0 & \frac{3}{4} & 1 & -\frac{1}{4} & \frac{3}{2} & 0 & 0 & 24 \\ 0 & 6 & 0 & -1 & 4 & 1 & 0 & 24 \\ \hline 0 & -20 & 0 & 10 & 20 & 0 & 1 & 10{,}560 \end{array}\right]\end{array}$$

$$\begin{array}{c} \\ c \\ t \\ s \\ M \end{array}\begin{array}{c}\begin{array}{ccccccc} c & s & t & u & v & w & M \end{array}\\ \left[\begin{array}{ccccccc|c} 1 & 0 & 0 & \frac{5}{12} & -\frac{5}{3} & \frac{1}{12} & 0 & 98 \\ 0 & 0 & 1 & -\frac{1}{8} & 1 & -\frac{1}{8} & 0 & 21 \\ 0 & 1 & 0 & -\frac{1}{6} & \frac{2}{3} & \frac{1}{6} & 0 & 4 \\ \hline 0 & 0 & 0 & \frac{20}{3} & \frac{100}{3} & \frac{10}{3} & 1 & 10{,}640 \end{array}\right]\end{array}$$

98 chairs, 4 sofas, 21 tables

17. Let b be the number of hours spent bicycling, j the number spent jogging, and s the number swimming each month. Maximize $200b + 475j + 275s$ subject to the

$$\text{constraints}\begin{cases} b + j + s \le 30 \\ s \le 4 \\ -b + j - s \le 0 \\ b \ge 0,\ j \ge 0,\ s \ge 0 \end{cases}$$

$$\begin{array}{c} \\ u \\ v \\ w \\ M \end{array}\begin{array}{c}\begin{array}{ccccccc} b & j & s & u & v & w & M \end{array}\\ \left[\begin{array}{ccccccc|c} 1 & 1 & 1 & 1 & 0 & 0 & 0 & 30 \\ 0 & 0 & 1 & 0 & 1 & 0 & 0 & 4 \\ -1 & 1 & -1 & 0 & 0 & 1 & 0 & 0 \\ \hline -200 & -475 & -275 & 0 & 0 & 0 & 1 & 0 \end{array}\right]\end{array}$$

$$\begin{array}{c} \\ u \\ v \\ j \\ M \end{array}\begin{array}{c}\begin{array}{ccccccc} b & j & s & u & v & w & M \end{array}\\ \left[\begin{array}{ccccccc|c} 2 & 0 & 2 & 1 & 0 & -1 & 0 & 30 \\ 0 & 0 & 1 & 0 & 1 & 0 & 0 & 4 \\ -1 & 1 & -1 & 0 & 0 & 1 & 0 & 0 \\ \hline -675 & 0 & -750 & 0 & 0 & 475 & 1 & 0 \end{array}\right]\end{array}$$

$$\begin{array}{c} \\ u \\ s \\ j \\ M \end{array}\begin{array}{c}\begin{array}{ccccccc} b & j & s & u & v & w & M \end{array}\\ \left[\begin{array}{ccccccc|c} 2 & 0 & 0 & 1 & -2 & -1 & 0 & 22 \\ 0 & 0 & 1 & 0 & 1 & 0 & 0 & 4 \\ -1 & 1 & 0 & 0 & 1 & 1 & 0 & 4 \\ \hline -675 & 0 & 0 & 0 & 750 & 475 & 1 & 3000 \end{array}\right]\end{array}$$

$$\begin{array}{c} \\ b \\ s \\ j \\ M \end{array}\begin{array}{c}\begin{array}{ccccccc} b & j & s & u & v & w & M \end{array}\\ \left[\begin{array}{ccccccc|c} 1 & 0 & 0 & \frac{1}{2} & -1 & -\frac{1}{2} & 0 & 11 \\ 0 & 0 & 1 & 0 & 1 & 0 & 0 & 4 \\ 0 & 1 & 0 & \frac{1}{2} & 0 & \frac{1}{2} & 0 & 15 \\ \hline 0 & 0 & 0 & \frac{675}{2} & 75 & \frac{275}{2} & 1 & 10{,}425 \end{array}\right]\end{array}$$

11 hours bicycling, 4 hours swimming, 15 hours jogging

He will lose $\dfrac{10{,}425}{3500} \approx 3$ pounds.

19. Let a be the number of type A restaurants, b the number of type B restaurants, and c the number of type C restaurants. Maximize $40a + 30b + 25c$ subject to the constraints

$$\begin{cases} 600a + 400b + 300c \le 48,000 \\ \quad 15a + 9b + 5c \le 1000 \\ \qquad a + b + c \le 70 \\ \qquad a \ge 0,\ b \ge 0,\ c \ge 0 \end{cases}$$

(dollars in thousands).

	a	b	c	u	v	w	M	
u	600	400	300	1	0	0	0	48,000
v	15	9	5	0	1	0	0	1000
w	1	1	1	0	0	1	0	70
M	−40	−30	−25	0	0	0	1	0

	a	b	c	u	v	w	M	
u	0	40	100	1	−40	0	0	8000
a	1	$\frac{3}{5}$	$\frac{1}{3}$	0	$\frac{1}{15}$	0	0	$\frac{200}{3}$
w	0	$\frac{2}{5}$	$\frac{2}{3}$	0	$-\frac{1}{15}$	1	0	$\frac{10}{3}$
M	0	−6	$-\frac{35}{3}$	0	$\frac{8}{3}$	0	1	$\frac{8000}{3}$

	a	b	c	u	v	w	M	
u	0	−20	0	1	−30	−150	0	7500
a	1	$\frac{2}{5}$	0	0	$\frac{1}{10}$	$-\frac{1}{2}$	0	65
c	0	$\frac{3}{5}$	1	0	$-\frac{1}{10}$	$\frac{3}{2}$	0	5
M	0	1	0	0	$\frac{3}{2}$	$\frac{35}{2}$	1	2725

65 type A restaurants and 5 type C restaurants

21.

	x	y	u	v	M	
u	1	4	1	0	0	300
v	1	2	0	1	0	200
M	−200	−500	0	0	1	0

	x	y	u	v	M	
y	$\frac{1}{4}$	1	$\frac{1}{4}$	0	0	75
v	$\frac{1}{2}$	0	$-\frac{1}{2}$	1	0	50
M	−75	0	125	0	1	37,500

	x	y	u	v	M	
y	0	1	$\frac{1}{2}$	$-\frac{1}{2}$	0	50
x	1	0	−1	2	0	100
M	0	0	50	150	1	45,000

$x = 100,\ y = 50;\ M = 45,000$

23.

	x	y	u	v	M	
u	1	4	1	0	0	4
v	3	2	0	1	0	6
M	−4	−6	0	0	1	0

	x	y	u	v	M	
y	$\frac{1}{4}$	1	$\frac{1}{4}$	0	0	1
v	$\frac{5}{2}$	0	$-\frac{1}{2}$	1	0	4
M	$-\frac{5}{2}$	0	$\frac{3}{2}$	0	1	6

	x	y	u	v	M	
y	0	1	$\frac{3}{10}$	$-\frac{1}{10}$	0	$\frac{3}{5}$
x	1	0	$-\frac{1}{5}$	$\frac{2}{5}$	0	$\frac{8}{5}$
M	0	0	1	1	1	10

$x = \dfrac{8}{5},\ y = \dfrac{3}{5};\ M = 10$

25.

$$\begin{array}{c}\ \\ u \\ v \\ w \\ M \end{array}\begin{array}{c}\begin{array}{ccccccc} x & y & z & u & v & w & M \end{array}\\ \left[\begin{array}{ccccccc|c} \underline{4} & 1 & 5 & 1 & 0 & 0 & 0 & 20 \\ 1 & 2 & 4 & 0 & 1 & 0 & 0 & 50 \\ 4 & 10 & 1 & 0 & 0 & 1 & 0 & 32 \\ \hline -16 & -4 & 20 & 0 & 0 & 0 & 1 & 0 \end{array}\right]\end{array}$$

$$\begin{array}{c}\ \\ x \\ v \\ w \\ M \end{array}\begin{array}{c}\begin{array}{ccccccc} x & y & z & u & v & w & M \end{array}\\ \left[\begin{array}{ccccccc|c} 1 & \frac{1}{4} & \frac{5}{4} & \frac{1}{4} & 0 & 0 & 0 & 5 \\ 0 & \frac{7}{4} & \frac{11}{4} & -\frac{1}{4} & 1 & 0 & 0 & 45 \\ 0 & 9 & -4 & -1 & 0 & 1 & 0 & 12 \\ \hline 0 & 0 & 40 & 4 & 0 & 0 & 1 & 80 \end{array}\right]\end{array}$$

$x = 5$, $y = 0$, $z = 0$; $M = 80$

Exercises 3

1.

$$\begin{array}{c}\ \\ u \\ v \\ M \end{array}\begin{array}{c}\begin{array}{ccccc} x & y & u & v & M \end{array}\\ \left[\begin{array}{ccccc|c} 1 & 1 & 1 & 0 & 0 & 5 \\ 2 & \underline{-3} & 0 & 1 & 0 & -12 \\ \hline -40 & -30 & 0 & 0 & 1 & 0 \end{array}\right]\end{array}$$

$$\begin{array}{c}\ \\ u \\ y \\ M \end{array}\begin{array}{c}\begin{array}{ccccc} x & y & u & v & M \end{array}\\ \left[\begin{array}{ccccc|c} \frac{5}{3} & 0 & 1 & \frac{1}{3} & 0 & 1 \\ -\frac{2}{3} & 1 & 0 & -\frac{1}{3} & 0 & 4 \\ \hline -60 & 0 & 0 & -10 & 1 & 120 \end{array}\right]\end{array}$$

$$\begin{array}{c}\ \\ x \\ y \\ M \end{array}\begin{array}{c}\begin{array}{ccccc} x & y & u & v & M \end{array}\\ \left[\begin{array}{ccccc|c} 1 & 0 & \frac{3}{5} & \frac{1}{5} & 0 & \frac{3}{5} \\ 0 & 1 & \frac{2}{5} & -\frac{1}{5} & 0 & \frac{22}{5} \\ \hline 0 & 0 & 36 & 2 & 1 & 156 \end{array}\right]\end{array}$$

$x = \dfrac{3}{5}$, $y = \dfrac{22}{5}$; $M = 156$

3.

$$\begin{array}{c}\ \\ u \\ v \\ M \end{array}\begin{array}{c}\begin{array}{ccccc} x & y & u & v & M \end{array}\\ \left[\begin{array}{ccccc|c} 1 & 1 & 1 & 0 & 0 & 3 \\ \underline{-2} & 0 & 0 & 1 & 0 & -5 \\ \hline 3 & 1 & 0 & 0 & 1 & 0 \end{array}\right]\end{array}$$

$$\begin{array}{c}\ \\ y \\ v \\ M \end{array}\begin{array}{c}\begin{array}{ccccc} x & y & u & v & M \end{array}\\ \left[\begin{array}{ccccc|c} 1 & 1 & -1 & 0 & 0 & 3 \\ \underline{-2} & 0 & 0 & 1 & 0 & -5 \\ \hline 2 & 0 & 1 & 0 & 1 & -3 \end{array}\right]\end{array}$$

$$\begin{array}{c}\ \\ y \\ x \\ M \end{array}\begin{array}{c}\begin{array}{ccccc} x & y & u & v & M \end{array}\\ \left[\begin{array}{ccccc|c} 0 & 1 & -1 & \frac{1}{2} & 0 & \frac{1}{2} \\ 1 & 0 & 0 & -\frac{1}{2} & 0 & \frac{5}{2} \\ \hline 0 & 0 & 1 & 1 & 1 & -8 \end{array}\right]\end{array}$$

$x = \dfrac{5}{2}$, $y = \dfrac{1}{2}$

Since $M = -8$, the minimum is 8. Pivoting about the -2 at the start leads to the same final matrix.

5.

$$\begin{array}{c}\ \\ u \\ v \\ w \\ t \\ M \end{array}\begin{array}{c}\begin{array}{ccccccc} x & y & u & v & w & t & M \end{array}\\ \left[\begin{array}{ccccccc|c} -2 & -1 & 1 & 0 & 0 & 0 & 0 & -11 \\ 1 & 1 & 0 & 1 & 0 & 0 & 0 & 10 \\ \frac{1}{3} & 1 & 0 & 0 & 1 & 0 & 0 & 6 \\ -\frac{1}{4} & \underline{-1} & 0 & 0 & 0 & 1 & 0 & -4 \\ \hline 13 & 4 & 0 & 0 & 0 & 0 & 1 & 0 \end{array}\right]\end{array}$$

$$\begin{array}{c}\ \\ u \\ v \\ w \\ y \\ M \end{array}\begin{array}{c}\begin{array}{ccccccc} x & y & u & v & w & t & M \end{array}\\ \left[\begin{array}{ccccccc|c} -\frac{7}{4} & 0 & 1 & 0 & 0 & -1 & 0 & -7 \\ \frac{3}{4} & 0 & 0 & 1 & 0 & 1 & 0 & 6 \\ \frac{1}{12} & 0 & 0 & 0 & 1 & 1 & 0 & 2 \\ \frac{1}{4} & 1 & 0 & 0 & 0 & -1 & 0 & 4 \\ \hline 12 & 0 & 0 & 0 & 0 & 4 & 1 & -16 \end{array}\right]\end{array}$$

$$\begin{array}{c}\ \\ u \\ v \\ t \\ y \\ M \end{array}\begin{array}{c}\begin{array}{ccccccc} x & y & u & v & w & t & M \end{array}\\ \left[\begin{array}{ccccccc|c} -\frac{5}{3} & 0 & 1 & 0 & 1 & 0 & 0 & -5 \\ \frac{2}{3} & 0 & 0 & 1 & -1 & 0 & 0 & 4 \\ \frac{1}{12} & 0 & 0 & 0 & 1 & 1 & 0 & 2 \\ \frac{1}{3} & 1 & 0 & 0 & 1 & 0 & 0 & 6 \\ \hline \frac{35}{3} & 0 & 0 & 0 & -4 & 0 & 1 & -24 \end{array}\right]\end{array}$$

$$\begin{array}{c}\ \\ x \\ v \\ t \\ y \\ M \end{array}\begin{array}{c}\begin{array}{ccccccc} x & y & u & v & w & t & M \end{array}\\ \left[\begin{array}{ccccccc|c} 1 & 0 & -\frac{3}{5} & 0 & -\frac{3}{5} & 0 & 0 & 3 \\ 0 & 0 & \frac{2}{5} & 1 & -\frac{3}{5} & 0 & 0 & 2 \\ 0 & 0 & \frac{1}{20} & 0 & \frac{21}{20} & 1 & 0 & \frac{7}{4} \\ 0 & 1 & \frac{1}{5} & 0 & \frac{6}{5} & 0 & 0 & 5 \\ \hline 0 & 0 & 7 & 0 & 3 & 0 & 1 & -59 \end{array}\right]\end{array}$$

$x = 3$, $y = 5$; since $M = -59$, the minimum is 59. Other choices of pivot entries lead to the same final matrix.

7. Minimize $3a + 1.5b$ subject to the constraints $\begin{cases} 30a + 10b \geq 60 \\ 10a + 10b \geq 40 \\ 20a + 60b \geq 120 \\ \quad\quad a \geq 0,\, b \geq 0 \end{cases}$

$$
\begin{array}{c}
\quad\;\; a \quad\;\; b \quad\; u \;\; v \;\; w \;\; M \\
\begin{array}{c} u \\ v \\ w \\ M \end{array}
\left[
\begin{array}{cccccc|c}
\underline{-30} & -10 & 1 & 0 & 0 & 0 & -60 \\
-10 & -10 & 0 & 1 & 0 & 0 & -40 \\
-20 & -60 & 0 & 0 & 1 & 0 & -120 \\
\hline
3 & \frac{3}{2} & 0 & 0 & 0 & 1 & 0
\end{array}
\right]
\end{array}
$$

$$
\begin{array}{c}
\quad\;\; a \quad\;\;\; b \quad\;\; u \;\; v \;\; w \;\; M \\
\begin{array}{c} a \\ v \\ w \\ M \end{array}
\left[
\begin{array}{cccccc|c}
1 & \frac{1}{3} & -\frac{1}{30} & 0 & 0 & 0 & 2 \\
0 & -\frac{20}{3} & -\frac{1}{3} & 1 & 0 & 0 & -20 \\
0 & -\frac{160}{3} & -\frac{2}{3} & 0 & 1 & 0 & -80 \\
\hline
0 & \frac{1}{2} & \frac{1}{10} & 0 & 0 & 1 & -6
\end{array}
\right]
\end{array}
$$

$$
\begin{array}{c}
\quad\;\; a \;\; b \quad\;\; u \quad\; v \quad\;\; w \;\; M \\
\begin{array}{c} a \\ v \\ b \\ M \end{array}
\left[
\begin{array}{cccccc|c}
1 & 0 & -\frac{3}{80} & 0 & \frac{1}{160} & 0 & \frac{3}{2} \\
0 & 0 & -\frac{1}{4} & 1 & -\frac{1}{8} & 0 & -10 \\
0 & 1 & \frac{1}{80} & 0 & -\frac{3}{160} & 0 & \frac{3}{2} \\
\hline
0 & 0 & \frac{3}{32} & 0 & \frac{3}{320} & 1 & -\frac{27}{4}
\end{array}
\right]
\end{array}
$$

$$
\begin{array}{c}
\quad\;\; a \;\; b \quad\; u \quad\;\; v \;\; w \;\; M \\
\begin{array}{c} a \\ w \\ b \\ M \end{array}
\left[
\begin{array}{cccccc|c}
1 & 0 & -\frac{1}{20} & \frac{1}{20} & 0 & 0 & 1 \\
0 & 0 & 2 & -8 & 1 & 0 & 80 \\
0 & 1 & \frac{1}{20} & -\frac{3}{20} & 0 & 0 & 3 \\
\hline
0 & 0 & \frac{3}{40} & \frac{3}{40} & 0 & 1 & -\frac{15}{2}
\end{array}
\right]
\end{array}
$$

1 serving of food A, 3 servings of food B

9. Maximize $30a + 50b + 60c$ subject to the constraints $\begin{cases} a + b + c \leq 600 \\ \quad\quad a \geq 100 \\ \quad\quad b \geq 50 \\ \quad b + c \geq 200 \\ a \geq 0,\, b \geq 0,\, c \geq 0 \end{cases}$

$$
\begin{array}{c}
\quad\;\; a \quad\; b \quad\;\; c \;\; u \;\; v \;\; w \;\; t \;\; M \\
\begin{array}{c} u \\ v \\ w \\ t \\ M \end{array}
\left[
\begin{array}{cccccccc|c}
1 & 1 & 1 & 1 & 0 & 0 & 0 & 0 & 600 \\
\underline{-1} & 0 & 0 & 0 & 1 & 0 & 0 & 0 & -100 \\
0 & -1 & 0 & 0 & 0 & 1 & 0 & 0 & -50 \\
0 & -1 & -1 & 0 & 0 & 0 & 1 & 0 & -200 \\
\hline
-30 & -50 & -60 & 0 & 0 & 0 & 0 & 1 & 0
\end{array}
\right]
\end{array}
$$

$$\begin{array}{c c} & \begin{array}{c c c c c c c c} a & b & c & u & v & w & t & M \end{array} \\ \begin{array}{c} u \\ a \\ w \\ t \\ M \end{array} & \left[\begin{array}{c c c c c c c c|c} 0 & 1 & 1 & 1 & 1 & 0 & 0 & 0 & 500 \\ 1 & 0 & 0 & 0 & -1 & 0 & 0 & 0 & 100 \\ 0 & \underline{-1} & 0 & 0 & 0 & 1 & 0 & 0 & -50 \\ 0 & -1 & -1 & 0 & 0 & 0 & 1 & 0 & -200 \\ \hline 0 & -50 & -60 & 0 & -30 & 0 & 0 & 1 & 3000 \end{array}\right] \end{array}$$

$$\begin{array}{c c} & \begin{array}{c c c c c c c c} a & b & c & u & v & w & t & M \end{array} \\ \begin{array}{c} u \\ a \\ b \\ t \\ M \end{array} & \left[\begin{array}{c c c c c c c c|c} 0 & 0 & 1 & 1 & 1 & 1 & 0 & 0 & 450 \\ 1 & 0 & 0 & 0 & -1 & 0 & 0 & 0 & 100 \\ 0 & 1 & 0 & 0 & 0 & -1 & 0 & 0 & 50 \\ 0 & 0 & \underline{-1} & 0 & 0 & -1 & 1 & 0 & -150 \\ \hline 0 & 0 & -60 & 0 & -30 & -50 & 0 & 1 & 5500 \end{array}\right] \end{array}$$

$$\begin{array}{c c} & \begin{array}{c c c c c c c c} a & b & c & u & v & w & t & M \end{array} \\ \begin{array}{c} u \\ a \\ b \\ c \\ M \end{array} & \left[\begin{array}{c c c c c c c c|c} 0 & 0 & 0 & 1 & 1 & 0 & \underline{1} & 0 & 300 \\ 1 & 0 & 0 & 0 & -1 & 0 & 0 & 0 & 100 \\ 0 & 1 & 0 & 0 & 0 & -1 & 0 & 0 & 50 \\ 0 & 0 & 1 & 0 & 0 & 1 & -1 & 0 & 150 \\ \hline 0 & 0 & 0 & 0 & -30 & 10 & -60 & 1 & 14,500 \end{array}\right] \end{array}$$

$$\begin{array}{c c} & \begin{array}{c c c c c c c c} a & b & c & u & v & w & t & M \end{array} \\ \begin{array}{c} t \\ a \\ b \\ c \\ M \end{array} & \left[\begin{array}{c c c c c c c c|c} 0 & 0 & 0 & 1 & 1 & 0 & 1 & 0 & 300 \\ 1 & 0 & 0 & 0 & -1 & 0 & 0 & 0 & 100 \\ 0 & 1 & 0 & 0 & 0 & -1 & 0 & 0 & 50 \\ 0 & 0 & 1 & 1 & 1 & 1 & 0 & 0 & 450 \\ \hline 0 & 0 & 0 & 60 & 30 & 10 & 0 & 1 & 32,500 \end{array}\right] \end{array}$$

Stock 100 of brand A, 50 of brand B, and 450 of brand C.

11.
$$\begin{array}{c c} & \begin{array}{c c c c c} x & y & u & v & M \end{array} \\ \begin{array}{c} u \\ v \\ M \end{array} & \left[\begin{array}{c c c c c|c} \underline{4} & 1 & 1 & 0 & 0 & 5 \\ -1 & -3 & 0 & 1 & 0 & -4 \\ \hline -1 & 2 & 0 & 0 & 1 & 0 \end{array}\right] \end{array}$$

$$\begin{array}{c c} & \begin{array}{c c c c c} x & y & u & v & M \end{array} \\ \begin{array}{c} x \\ v \\ M \end{array} & \left[\begin{array}{c c c c c|c} 1 & \frac{1}{4} & \frac{1}{4} & 0 & 0 & \frac{5}{4} \\ 0 & -\frac{11}{4} & \frac{1}{4} & 1 & 0 & -\frac{11}{4} \\ \hline 0 & \frac{9}{4} & \frac{1}{4} & 0 & 1 & \frac{5}{4} \end{array}\right] \end{array}$$

$$\begin{array}{c} & \begin{array}{ccccc} x & y & u & v & M \end{array} \\ \begin{array}{c} x \\ y \\ M \end{array} & \left[\begin{array}{ccccc|c} 1 & 0 & \frac{3}{11} & \frac{1}{11} & 0 & 1 \\ 0 & 1 & -\frac{1}{11} & -\frac{4}{11} & 0 & 1 \\ \hline 0 & 0 & \frac{5}{11} & \frac{9}{11} & 1 & -1 \end{array}\right] \end{array}$$

$x = 1, y = 1; M = -1$

Exercises 4

1. x (paring knives) goes from 14 to

$14 + 54\left(-\dfrac{5}{27}\right) = 4$, y (pocket knives) goes

from 8 to $8 + 54\left(\dfrac{7}{27}\right) = 22$, and the profit

goes from $82 to $82 + 54\left(\dfrac{20}{27}\right) = \$122$.

3. The slack variable involved is w. x (sets to Rockville) goes from 20 to
$20 + (50 - 45)(1) = 25$, y (sets to Annapolis) goes from 25 to $25 + (50 - 45)(0) = 25$, and the cost goes from $260 to
$-[-260 + (50 - 45)(2)] = \$250$.

5. $14 + h\left(\dfrac{1}{3}\right) \geq 0$, $4 + h\left(-\dfrac{1}{3}\right) \geq 0$, and

$20 + h\left(\dfrac{4}{3}\right) \geq 0$, so $h \geq -42$, $h \leq 12$, and

$h \geq -15$, or $-15 \leq h \leq 12$.

7. $\begin{bmatrix} 9 & 1 & 1 \\ 4 & 8 & -3 \end{bmatrix}$

9. $\begin{bmatrix} 7 \\ 6 \\ 5 \\ 1 \end{bmatrix}$

11. Yes

13. Minimize $[7 \quad 5 \quad 4]\begin{bmatrix} x \\ y \\ z \end{bmatrix}$ subject to the

constraints $\begin{bmatrix} 3 & 8 & 9 \\ 1 & 2 & 5 \\ 4 & 1 & 7 \end{bmatrix}\begin{bmatrix} x \\ y \\ z \end{bmatrix} \geq \begin{bmatrix} 75 \\ 80 \\ 67 \end{bmatrix}$ and

$\begin{bmatrix} x \\ y \\ z \end{bmatrix} \geq \begin{bmatrix} 0 \\ 0 \\ 0 \end{bmatrix}$.

15. Maximize $[3 \quad 5]\begin{bmatrix} x \\ y \end{bmatrix}$ subject to the

constraints $\begin{bmatrix} 3 & 6 \\ 7 & 5 \\ 4 & 3 \end{bmatrix}\begin{bmatrix} x \\ y \end{bmatrix} \leq \begin{bmatrix} 90 \\ 138 \\ 120 \end{bmatrix}$ and

$\begin{bmatrix} x \\ y \end{bmatrix} \leq \begin{bmatrix} 0 \\ 0 \end{bmatrix}$.

17. Minimize $2x + 3y$ subject to the constraints
$$\begin{cases} 7x + 4y \geq 33 \\ 5x + 8y \geq 44 \\ x + 3y \geq 55 \\ x \geq 0, \ y \geq 0 \end{cases}$$

Exercises 5

1. Primal: Maximize $[4 \quad 2]\begin{bmatrix} x \\ y \end{bmatrix}$ subject to

$\begin{bmatrix} 5 & 1 \\ 3 & 2 \end{bmatrix}\begin{bmatrix} x \\ y \end{bmatrix} \leq \begin{bmatrix} 80 \\ 76 \end{bmatrix}$ and $\begin{bmatrix} x \\ y \end{bmatrix} \geq \begin{bmatrix} 0 \\ 0 \end{bmatrix}$.

Dual: Minimize $[80 \quad 76]\begin{bmatrix} u \\ v \end{bmatrix}$ subject to

$\begin{bmatrix} 5 & 3 \\ 1 & 2 \end{bmatrix}\begin{bmatrix} u \\ v \end{bmatrix} \geq \begin{bmatrix} 4 \\ 2 \end{bmatrix}$ and $\begin{bmatrix} u \\ v \end{bmatrix} \geq \begin{bmatrix} 0 \\ 0 \end{bmatrix}$.

Minimize $80u + 76v$ subject to the

constraints $\begin{cases} 5u + 3v \geq 4 \\ u + 2v \geq 2 \\ u \geq 0, \ v \geq 0 \end{cases}$.

3. Primal: Minimize $[10 \quad 12]\begin{bmatrix} x \\ y \end{bmatrix}$ subject to

$$\begin{bmatrix} 1 & 2 \\ -1 & 1 \\ 2 & 3 \end{bmatrix}\begin{bmatrix} x \\ y \end{bmatrix} \geq \begin{bmatrix} 1 \\ 2 \\ 1 \end{bmatrix} \text{ and } \begin{bmatrix} x \\ y \end{bmatrix} \geq \begin{bmatrix} 0 \\ 0 \end{bmatrix}.$$

Dual: Maximize $[1 \quad 2 \quad 1]\begin{bmatrix} u \\ v \\ w \end{bmatrix}$ subject to

$$\begin{bmatrix} 1 & -1 & 2 \\ 2 & 1 & 3 \end{bmatrix}\begin{bmatrix} u \\ v \\ w \end{bmatrix} \leq \begin{bmatrix} 10 \\ 12 \end{bmatrix} \text{ and } \begin{bmatrix} u \\ v \\ w \end{bmatrix} \geq \begin{bmatrix} 0 \\ 0 \\ 0 \end{bmatrix}.$$

Maximize $u + 2v + w$ subject to the

constraints $\begin{cases} u - v + 2w \leq 10 \\ 2u + v + 3w \leq 12 \\ \quad\quad u \geq 0, v \geq 0, w \geq 0 \end{cases}$

5. Primal: Minimize $[3 \quad 5 \quad 1]\begin{bmatrix} x \\ y \\ z \end{bmatrix}$ subject to

$$\begin{bmatrix} -2 & 4 & 6 \\ 8 & 1 & 9 \end{bmatrix}\begin{bmatrix} x \\ y \\ z \end{bmatrix} \geq \begin{bmatrix} -7 \\ 10 \end{bmatrix} \text{ and } \begin{bmatrix} x \\ y \\ z \end{bmatrix} \geq \begin{bmatrix} 0 \\ 0 \\ 0 \end{bmatrix}.$$

Dual: Maximize $[-7 \quad 10]\begin{bmatrix} u \\ v \end{bmatrix}$ subject to

$$\begin{bmatrix} -2 & 8 \\ 4 & 1 \\ 6 & 9 \end{bmatrix}\begin{bmatrix} u \\ v \end{bmatrix} \leq \begin{bmatrix} 3 \\ 5 \\ 1 \end{bmatrix} \text{ and } \begin{bmatrix} u \\ v \end{bmatrix} \geq \begin{bmatrix} 0 \\ 0 \end{bmatrix}.$$

Maximize $-7u + 10v$ subject to the

constraints $\begin{cases} -2u + 8v \leq 3 \\ 4u + v \leq 5 \\ 6u + 9v \leq 1 \\ \quad\quad u \geq 0, v \geq 0 \end{cases}$

7. $x = 12, y = 20, M = 88$; $u = \dfrac{2}{7}$, $u = \dfrac{6}{7}$,

$M = 88$

9. $x = 0, y = 2, M = 24$; $u = 0, v = 12, M = 24$

11. Maximize $3u + 5v$ subject to the constraints

$\begin{cases} u + 2v \leq 3 \\ \quad 2u \leq 1 \\ \quad\quad u \geq 0, v \geq 0 \end{cases}$

Solve the dual.

$$\begin{array}{c}\begin{array}{cccccc} u & v & x & y & M \end{array} \\ \begin{array}{c} x \\ y \\ M \end{array}\left[\begin{array}{ccccc|c} 1 & \underline{2} & 1 & 0 & 0 & 3 \\ 1 & 0 & 0 & 1 & 0 & 1 \\ \hline -3 & -5 & 0 & 0 & 1 & 0 \end{array}\right]\end{array}$$

$$\begin{array}{c}\begin{array}{ccccc} u & v & x & y & M \end{array} \\ \begin{array}{c} v \\ y \\ M \end{array}\left[\begin{array}{ccccc|c} \frac{1}{2} & 1 & \frac{1}{2} & 0 & 0 & \frac{3}{2} \\ \underline{1} & 0 & 0 & 1 & 0 & 1 \\ \hline -\frac{1}{2} & 0 & \frac{5}{2} & 0 & 1 & \frac{15}{2} \end{array}\right]\end{array}$$

$$\begin{array}{c}\begin{array}{ccccc} u & v & x & y & M \end{array} \\ \begin{array}{c} v \\ u \\ M \end{array}\left[\begin{array}{ccccc|c} 0 & 1 & \frac{1}{2} & -\frac{1}{2} & 0 & 1 \\ 1 & 0 & 0 & 1 & 0 & 1 \\ \hline 0 & 0 & \frac{5}{2} & \frac{1}{2} & 1 & 8 \end{array}\right]\end{array}$$

$x = \dfrac{5}{2}$, $y = \dfrac{1}{2}$, minimum $= 8$

$u = 1, v = 1$, maximum $= 8$

13. Minimize $6u + 9v + 12w$ subject to the

constraints $\begin{cases} u + 3v \geq 10 \\ 2u - w \geq 12 \\ v + 3w \geq 10 \\ u \geq 0, v \geq 0, w \geq 0 \end{cases}$

Solve the primal.

$$\begin{array}{c}\begin{array}{ccccccc} x & y & z & u & v & w & M \end{array} \\ \begin{array}{c} u \\ v \\ w \\ M \end{array}\left[\begin{array}{ccccccc|c} 1 & -2 & 0 & 1 & 0 & 0 & 0 & 6 \\ 3 & 0 & 1 & 0 & 1 & 0 & 0 & 9 \\ 0 & \underline{1} & 3 & 0 & 0 & 1 & 0 & 12 \\ \hline -10 & -12 & -10 & 0 & 0 & 0 & 1 & 0 \end{array}\right]\end{array}$$

$$\begin{array}{c}\begin{array}{ccccccc} x & y & z & u & v & w & M \end{array} \\ \begin{array}{c} u \\ v \\ y \\ M \end{array}\left[\begin{array}{ccccccc|c} 1 & 0 & 6 & 1 & 0 & 2 & 0 & 30 \\ \underline{3} & 0 & 1 & 0 & 1 & 0 & 0 & 9 \\ 0 & 1 & 3 & 0 & 0 & 1 & 0 & 12 \\ \hline -10 & 0 & 26 & 0 & 0 & 12 & 1 & 144 \end{array}\right]\end{array}$$

$$\begin{array}{c} \\ u \\ x \\ y \\ M \end{array} \begin{array}{cccccc} x & y & z & u & v & w & M \\ \left[\begin{array}{ccccccc|c} 0 & 0 & \frac{17}{3} & 1 & -\frac{1}{3} & 2 & 0 & 27 \\ 1 & 0 & \frac{1}{3} & 0 & \frac{1}{3} & 0 & 0 & 3 \\ 0 & 1 & 3 & 0 & 0 & 1 & 0 & 12 \\ 0 & 0 & \frac{88}{3} & 0 & \frac{10}{3} & 12 & 1 & 174 \end{array}\right] \end{array}$$

$x = 3$, $y = 12$, $z = 0$, maximum = 174

$u = 0$, $v = \dfrac{10}{3}$, $w = 12$, minimum = 174

15. Suppose we can hire workers out at a profit of u dollars per hour, sell the steel at a profit of v dollars per unit, and sell the wood at a profit of w dollars per unit. To find the minimum profit at which that should be done, minimize $90u + 138v + 120w$ subject to the constraints

$$\begin{cases} 3u + 7v + 4w \geq 3 \\ 6u + 5v + 3w \geq 5 \\ \qquad u \geq 0,\ v \geq 0,\ w \geq 0 \end{cases}$$

17. Suppose we can buy anthracite at u dollars per ton, ordinary coal at v dollars per ton, and bituminous coal at w dollars per ton. To find the maximum cost at which this should be done, maximize $80u + 60v + 75w$ subject to the constraints

$$\begin{cases} 4u + 4v + 7w \leq 150 \\ 10u + 5v + 5w \leq 200 \\ \qquad u \geq 0,\ v \geq 0,\ w \geq 0 \end{cases}$$

19. The new primal problem is to maximize $3x + 5y + pz$, where p is the profit per table knife and z is the number of table knives produced, subject to the constraints

$$\begin{cases} 3x + 6y + 4z \leq 90 \\ 7x + 5y + 6z \leq 138 \\ 4x + 3y + 2z \leq 120 \\ \qquad x \geq 0,\ y \geq 0,\ z \geq 0 \end{cases}$$

The dual is to minimize $90u + 138v + 120w$ subject to the constraints

$$\begin{cases} 3u + 7v + 4w \geq 3 \\ 6u + 5v + 3w \geq 5 \\ 4u + 6v + 2w \geq p \\ \qquad u \geq 0,\ v \geq 0,\ w \geq 0 \end{cases}$$

The original solution, with $u = \dfrac{20}{7}$, $v = \dfrac{1}{9}$, $w = 0$, will still be optimal if

$4\left(\dfrac{20}{27}\right) + 6\left(\dfrac{1}{9}\right) + 2(0) \geq p$. Since

$4\left(\dfrac{20}{27}\right) + 6\left(\dfrac{1}{9}\right) + 2(0) \approx \$3.63$, that is the minimum profit per table knife that needs to be realized to warrant adding table knives to the product line.

21. Dual: Maximize $5u + 8v$ subject to the constraints $\begin{cases} u + 2v \leq 16 \\ 3u + 4v \leq 42 \\ \qquad u \geq 0,\ v \geq 0 \end{cases}$

$$\begin{array}{c} \\ x \\ y \\ M \end{array} \begin{array}{ccccc} u & v & x & y & M \\ \left[\begin{array}{ccccc|c} 1 & \underline{2} & 1 & 0 & 0 & 16 \\ 3 & 4 & 0 & 1 & 0 & 42 \\ -5 & -8 & 0 & 0 & 1 & 0 \end{array}\right] \end{array}$$

$$\begin{array}{c} \\ v \\ y \\ M \end{array} \begin{array}{ccccc} u & v & x & y & M \\ \left[\begin{array}{ccccc|c} \frac{1}{2} & 1 & \frac{1}{2} & 0 & 0 & 8 \\ \underline{1} & 0 & -2 & 1 & 0 & 10 \\ -1 & 0 & 4 & 0 & 1 & 64 \end{array}\right] \end{array}$$

$$\begin{array}{c} \\ v \\ u \\ M \end{array} \begin{array}{ccccc} u & v & x & y & M \\ \left[\begin{array}{ccccc|c} 0 & 1 & \frac{3}{2} & -\frac{1}{2} & 0 & 3 \\ 1 & 0 & -2 & 1 & 0 & 10 \\ 0 & 0 & 2 & 1 & 1 & 74 \end{array}\right] \end{array}$$

$y = 2$, $y = 1$, $M = 74$

Chapter 4 Supplementary Exercises

1.

	x	y	u	v	M	
u	2	1	1	0	0	7
v	−1	1	0	1	0	1
M	−3	−4	0	0	1	0

	x	y	u	v	M	
u	3	0	1	−1	0	6
y	−1	1	0	1	0	1
M	−7	0	0	4	1	4

	x	y	u	v	M	
x	1	0	$\frac{1}{3}$	$-\frac{1}{3}$	0	2
y	0	1	$\frac{1}{3}$	$\frac{2}{3}$	0	3
M	0	0	$\frac{7}{3}$	$\frac{5}{3}$	1	18

$x = 2, y = 3, M = 18$

2.

	x	y	u	v	M	
u	1	1	1	0	0	7
v	4	3	0	1	0	24
M	−2	−5	0	0	1	0

	x	y	u	v	M	
y	1	1	1	0	0	7
v	1	0	−3	1	0	3
M	3	0	5	0	1	35

$x = 0, y = 7, M = 35$

3.

	x	y	u	v	w	M	
u	1	2	1	0	0	0	14
v	1	1	0	1	0	0	9
w	3	2	0	0	1	0	24
M	−2	−3	0	0	0	1	0

	x	y	u	v	w	M	
y	$\frac{1}{2}$	1	$\frac{1}{2}$	0	0	0	7
v	$\frac{1}{2}$	0	$-\frac{1}{2}$	1	0	0	2
w	2	0	−1	0	1	0	10
M	$-\frac{1}{2}$	0	$\frac{3}{2}$	0	0	1	21

	x	y	u	v	w	M	
y	0	1	1	−1	0	0	5
x	1	0	−1	2	0	0	4
w	0	0	1	−4	1	0	2
M	0	0	1	1	0	1	23

$x = 4, y = 5, M = 23$

4.

	x	y	u	v	w	M	
u	1	2	1	0	0	0	10
v	4	3	0	1	0	0	30
w	−2	1	0	0	1	0	0
M	−3	−7	0	0	0	1	0

	x	y	u	v	w	M	
u	5	0	1	0	−2	0	10
v	10	0	0	1	−3	0	30
y	−2	1	0	0	1	0	0
M	−17	0	0	0	7	1	0

	x	y	u	v	w	M	
y	1	0	$\frac{1}{5}$	0	$-\frac{2}{5}$	0	2
v	0	0	−2	1	1	0	10
y	0	1	$\frac{2}{5}$	0	$\frac{1}{5}$	0	4
M	0	0	$\frac{17}{5}$	0	$\frac{1}{5}$	1	34

$x = 2, y = 4, M = 34$

5.

	x	y	u	v	M	
u	−7	−5	1	0	0	−40
v	−1	−4	0	1	0	−9
M	1	1	0	0	1	0

	x	y	u	v	M	
x	1	$\frac{5}{7}$	$-\frac{1}{7}$	0	0	$\frac{40}{7}$
v	0	$-\frac{23}{7}$	$-\frac{1}{7}$	1	0	$-\frac{23}{7}$
M	0	$\frac{2}{7}$	$\frac{1}{7}$	0	1	$-\frac{40}{7}$

	x	y	u	v	M	
x	1	0	$-\frac{4}{23}$	$\frac{5}{23}$	0	5
y	0	1	$\frac{1}{23}$	$-\frac{7}{23}$	0	1
M	0	0	$\frac{3}{23}$	$\frac{2}{23}$	1	−6

$x = 5, y = 1, \text{min.} = 6$

6.

$$\begin{array}{c c} & \begin{array}{ccccc} x & y & u & v & M \end{array} \\ \begin{array}{c} u \\ v \\ M \end{array} & \left[\begin{array}{ccccc|c} \underline{-1} & -1 & 1 & 0 & 0 & -6 \\ -1 & -2 & 0 & 1 & 0 & 0 \\ \hline 3 & 2 & 0 & 0 & 1 & 0 \end{array}\right] \end{array}$$

$$\begin{array}{c c} & \begin{array}{ccccc} x & y & u & v & M \end{array} \\ \begin{array}{c} x \\ v \\ M \end{array} & \left[\begin{array}{ccccc|c} 1 & \underline{1} & -1 & 0 & 0 & 6 \\ 0 & -1 & -1 & 1 & 0 & 6 \\ \hline 0 & -1 & 3 & 0 & 1 & -18 \end{array}\right] \end{array}$$

$$\begin{array}{c c} & \begin{array}{ccccc} x & y & u & v & M \end{array} \\ \begin{array}{c} y \\ v \\ M \end{array} & \left[\begin{array}{ccccc|c} 1 & 1 & -1 & 0 & 0 & 6 \\ 1 & 0 & -2 & 1 & 0 & 12 \\ \hline 1 & 0 & 2 & 0 & 1 & -12 \end{array}\right] \end{array}$$

$x = 0, y = 6, M = 12$
The minimum is 12.

7.

$$\begin{array}{c c} & \begin{array}{cccccc} x & y & u & v & w & M \end{array} \\ \begin{array}{c} u \\ v \\ w \\ M \end{array} & \left[\begin{array}{cccccc|c} -1 & -4 & 1 & 0 & 0 & 0 & -8 \\ -1 & -1 & 0 & 1 & 0 & 0 & -5 \\ \underline{-2} & -1 & 0 & 0 & 1 & 0 & -7 \\ \hline 20 & 30 & 0 & 0 & 0 & 1 & 0 \end{array}\right] \end{array}$$

$$\begin{array}{c c} & \begin{array}{cccccc} x & y & u & v & w & M \end{array} \\ \begin{array}{c} u \\ v \\ x \\ M \end{array} & \left[\begin{array}{cccccc|c} 0 & -\frac{7}{2} & 1 & 0 & -\frac{1}{2} & 0 & -\frac{9}{2} \\ 0 & -\frac{1}{2} & 0 & 1 & -\frac{1}{2} & 0 & -\frac{3}{2} \\ 1 & \frac{1}{2} & 0 & 0 & -\frac{1}{2} & 0 & \frac{7}{2} \\ \hline 0 & 20 & 0 & 0 & 10 & 1 & -70 \end{array}\right] \end{array}$$

$$\begin{array}{c c} & \begin{array}{cccccc} x & y & u & v & w & M \end{array} \\ \begin{array}{c} u \\ w \\ x \\ M \end{array} & \left[\begin{array}{cccccc|c} 0 & \underline{-3} & 1 & -1 & 0 & 0 & -3 \\ 0 & 1 & 0 & -2 & 1 & 0 & 3 \\ 1 & 1 & 0 & -1 & 0 & 0 & 5 \\ \hline 0 & 10 & 0 & 20 & 0 & 1 & -100 \end{array}\right] \end{array}$$

$$\begin{array}{c c} & \begin{array}{cccccc} x & y & u & v & w & M \end{array} \\ \begin{array}{c} y \\ w \\ x \\ M \end{array} & \left[\begin{array}{cccccc|c} 0 & 1 & -\frac{1}{3} & \frac{1}{3} & 0 & 0 & 1 \\ 0 & 0 & \frac{1}{3} & -\frac{7}{3} & 1 & 0 & 2 \\ 1 & 0 & \frac{1}{3} & -\frac{4}{3} & 0 & 0 & 4 \\ \hline 0 & 0 & \frac{10}{3} & \frac{50}{3} & 0 & 1 & -110 \end{array}\right] \end{array}$$

$x = 4, y = 1, M = -110$
The minimum is 110.

8.

$$\begin{array}{c c} & \begin{array}{cccccc} x & y & u & v & w & M \end{array} \\ \begin{array}{c} u \\ v \\ w \\ M \end{array} & \left[\begin{array}{cccccc|c} -2 & -1 & 1 & 0 & 0 & 0 & -10 \\ -3 & -2 & 0 & 1 & 0 & 0 & -18 \\ -1 & \underline{-2} & 0 & 0 & 1 & 0 & -10 \\ \hline 5 & 7 & 0 & 0 & 0 & 1 & 0 \end{array}\right] \end{array}$$

$$\begin{array}{c c} & \begin{array}{cccccc} x & y & u & v & w & M \end{array} \\ \begin{array}{c} u \\ v \\ y \\ M \end{array} & \left[\begin{array}{cccccc|c} -\frac{3}{2} & 0 & 1 & 0 & -\frac{1}{2} & 0 & -5 \\ -2 & 0 & 0 & 1 & \underline{-1} & 0 & -8 \\ \frac{1}{2} & 1 & 0 & 0 & -\frac{1}{2} & 0 & 5 \\ \hline \frac{3}{2} & 0 & 0 & 0 & \frac{7}{2} & 1 & -35 \end{array}\right] \end{array}$$

$$\begin{array}{c c} & \begin{array}{cccccc} x & y & u & v & w & M \end{array} \\ \begin{array}{c} u \\ w \\ y \\ M \end{array} & \left[\begin{array}{cccccc|c} -\frac{1}{2} & 0 & 1 & -\frac{1}{2} & 0 & 0 & -1 \\ 2 & 0 & 0 & -1 & 1 & 0 & 8 \\ \frac{3}{2} & 1 & 0 & -\frac{1}{2} & 0 & 0 & 9 \\ \hline -\frac{11}{2} & 0 & 0 & \frac{7}{2} & 0 & 1 & -63 \end{array}\right] \end{array}$$

$$\begin{array}{c c} & \begin{array}{cccccc} x & y & u & v & w & M \end{array} \\ \begin{array}{c} x \\ w \\ y \\ M \end{array} & \left[\begin{array}{cccccc|c} 1 & 0 & -2 & 1 & 0 & 0 & 2 \\ 0 & 0 & \underline{4} & -3 & 1 & 0 & 4 \\ 0 & 1 & 3 & -2 & 0 & 0 & 6 \\ \hline 0 & 0 & -11 & 9 & 0 & 1 & -52 \end{array}\right] \end{array}$$

$$\begin{array}{c c} & \begin{array}{cccccc} x & y & u & v & w & M \end{array} \\ \begin{array}{c} x \\ u \\ y \\ M \end{array} & \left[\begin{array}{cccccc|c} 1 & 0 & 0 & -\frac{1}{2} & \frac{1}{2} & 0 & 4 \\ 0 & 0 & 1 & -\frac{3}{4} & \frac{1}{4} & 0 & 1 \\ 0 & 1 & 0 & \frac{1}{4} & -\frac{3}{4} & 0 & 3 \\ \hline 0 & 0 & 0 & \frac{3}{4} & \frac{11}{4} & 1 & -41 \end{array}\right] \end{array}$$

$x = 4, y = 3, M = -41$
The minimum is 41.

9.

	x	y	z	t	u	v	w	M	
t	1	0	0	1	0	0	0	0	4
u	0	1	0	0	1	0	0	0	6
v	0	0	$\underline{1}$	0	0	1	0	0	8
w	4	3	2	0	0	0	1	0	38
M	−36	−48	−70	0	0	0	0	1	0

	x	y	z	t	u	v	w	M	
t	1	0	0	1	0	0	0	0	4
u	0	$\underline{1}$	0	0	1	0	0	0	6
z	0	0	1	0	0	1	0	0	8
w	4	3	0	0	0	−2	1	0	22
M	−36	−48	0	0	0	70	0	1	560

	x	y	z	t	u	v	w	M	
t	1	0	0	1	0	0	0	0	4
y	0	1	0	0	1	0	0	0	6
z	0	0	1	0	0	1	0	0	8
w	$\underline{4}$	0	0	0	−3	−2	1	0	4
M	−36	0	0	0	48	70	0	1	848

	x	y	z	t	u	v	w	M	
t	0	0	0	1	$\frac{3}{4}$	$\frac{1}{2}$	$-\frac{1}{4}$	0	3
y	0	1	0	0	1	0	0	0	6
z	0	0	1	0	0	1	0	0	8
x	1	0	0	0	$-\frac{3}{4}$	$-\frac{1}{2}$	$\frac{1}{4}$	0	1
M	0	0	0	0	21	52	9	1	884

$x = 1, y = 6, z = 8, M = 884$

10.

	x	y	z	w	t	u	v	M	
t	6	9	12	15	1	0	0	0	672
u	1	−1	$\underline{2}$	2	0	1	0	0	92
v	5	10	−5	4	0	0	1	0	280
M	−3	−4	−5	−4	0	0	0	1	0

	x	y	z	w	t	u	v	M	
t	0	$\underline{15}$	0	3	1	−6	0	0	120
z	$\frac{1}{2}$	$-\frac{1}{2}$	1	1	0	$\frac{1}{2}$	0	0	46
v	$\frac{15}{2}$	$\frac{15}{2}$	0	9	0	$\frac{5}{2}$	1	0	510
M	$-\frac{1}{2}$	$-\frac{13}{2}$	0	1	0	$\frac{5}{2}$	0	1	230

$$\begin{array}{c}\\ y\\ z\\ v\\ M\end{array}\begin{array}{c}\begin{array}{cccccccc}x & y & z & w & t & u & v & M\end{array}\\ \left[\begin{array}{cccccccc|c} 0 & 1 & 0 & \frac{1}{5} & \frac{1}{15} & -\frac{2}{5} & 0 & 0 & 8\\ \frac{1}{2} & 0 & 1 & \frac{11}{10} & \frac{1}{30} & \frac{3}{10} & 0 & 0 & 50\\ \frac{15}{2} & 0 & 0 & \frac{15}{2} & -\frac{1}{2} & \frac{11}{2} & 1 & 0 & 450\\ \hline -\frac{1}{2} & 0 & 0 & \frac{23}{10} & \frac{13}{30} & -\frac{1}{10} & 0 & 1 & 282\end{array}\right]\end{array}$$

$$\begin{array}{c}\\ y\\ z\\ x\\ M\end{array}\begin{array}{c}\begin{array}{cccccccc}x & y & z & w & t & u & v & M\end{array}\\ \left[\begin{array}{cccccccc|c} 0 & 1 & 0 & \frac{1}{5} & \frac{1}{15} & -\frac{2}{5} & 0 & 0 & 8\\ 0 & 0 & 1 & \frac{3}{5} & \frac{1}{15} & -\frac{1}{15} & -\frac{1}{15} & 0 & 20\\ 1 & 0 & 0 & 1 & -\frac{1}{15} & \frac{11}{15} & \frac{2}{15} & 0 & 60\\ \hline 0 & 0 & 0 & \frac{14}{5} & \frac{2}{5} & \frac{4}{15} & \frac{1}{15} & 1 & 312\end{array}\right]\end{array}$$

$x = 60,\ y = 8,\ z = 20,\ w = 0,\ M = 312$

11. Minimize $14u + 9v + 24w$ subject to the

constraints $\begin{cases} u + v + 3w \geq 2\\ 2u + v + 2w \geq 3\\ u \geq 0,\ v \geq 0,\ w \geq 0\end{cases}$

12. Maximize $8u + 5v + 7w$ subject to the

constraints $\begin{cases} u + v + 2w \leq 20\\ 4u + v + w \leq 30\\ u \geq 0,\ v \geq 0,\ w \geq 0\end{cases}$

13. Primal: $x = 4,\ y = 5,$ maximum $= 23$;
dual: $u = 1,\ v = 1,\ w = 0,$ minimum $= 23$

14. Primal: $x = 4,\ y = 1,$ minimum $= 110$;

dual: $u = \dfrac{10}{3},\ v = \dfrac{50}{3},\ w = 0,$

maximum $= 110$

15. $A = \begin{bmatrix} 1 & 2\\ 1 & 1\\ 3 & 2\end{bmatrix}$, $B = \begin{bmatrix} 14\\ 9\\ 24\end{bmatrix}$, $C = [2\ \ 3]$, $X = \begin{bmatrix} x\\ y\end{bmatrix}$

Primal: Maximize CX subject to $AX \leq B$,
$X \geq \mathbf{0}$.

Dual: $U = \begin{bmatrix} u\\ v\\ w\end{bmatrix}$

Minimize $B^T U$ subject to
$A^T U \geq C^T,\ U \geq \mathbf{0}$.

16. $A = \begin{bmatrix} 1 & 4\\ 1 & 1\\ 2 & 1\end{bmatrix}$, $B = \begin{bmatrix} 8\\ 5\\ 7\end{bmatrix}$, $C = [20\ \ 30]$,

$X = \begin{bmatrix} x\\ y\end{bmatrix}$

Minimize CX subject to $AX \geq B,\ X \geq \mathbf{0}$.

Dual: $U = \begin{bmatrix} u\\ v\\ w\end{bmatrix}$

Maximize $B^T U$ subject to $A^T U \leq C^T$,
$U \geq \mathbf{0}$.

17. a. Let a be the number of type A sticks
and b the number of type B sticks.
Maximize $8a + 10b$ subject to the

constraints $\begin{cases} 2a + b \leq 120\\ a + 3b \leq 150\\ 2a + 2b \leq 140\\ a \geq 0,\ b \geq 0\end{cases}$

$$\begin{array}{c}\\ u\\ v\\ w\\ M\end{array}\begin{array}{c}\begin{array}{cccccc}a & b & u & v & w & M\end{array}\\ \left[\begin{array}{cccccc|c} 2 & 1 & 1 & 0 & 0 & 0 & 120\\ 1 & 3 & 0 & 1 & 0 & 0 & 150\\ 2 & 2 & 0 & 0 & 1 & 0 & 140\\ \hline -8 & -10 & 0 & 0 & 0 & 1 & 0\end{array}\right]\end{array}$$

$$\begin{array}{c|cccccc|c} & a & b & u & v & w & M & \\ \hline u & \frac{5}{3} & 0 & 1 & -\frac{1}{3} & 0 & 0 & 70 \\ b & \frac{1}{3} & 1 & 0 & \frac{1}{3} & 0 & 0 & 50 \\ w & \frac{4}{3} & 0 & 0 & -\frac{2}{3} & 1 & 0 & 40 \\ \hline M & -\frac{14}{3} & 0 & 0 & \frac{10}{3} & 0 & 1 & 500 \end{array}$$

$$\begin{array}{c|cccccc|c} & a & b & u & v & w & M & \\ \hline u & 0 & 0 & 1 & \frac{1}{2} & -\frac{5}{4} & 0 & 20 \\ b & 0 & 1 & 0 & \frac{1}{2} & -\frac{1}{4} & 0 & 40 \\ a & 1 & 0 & 0 & -\frac{1}{2} & \frac{3}{4} & 0 & 30 \\ \hline M & 0 & 0 & 0 & 1 & \frac{7}{2} & 1 & 640 \end{array}$$

30 type A sticks, 40 type B sticks

b. The new problem is to maximize $8a + 10b + pc$, where c is the number of tennis rackets and p is the profit on each racket, subject to the constraints

$$\begin{cases} 2a + b + c \leq 120 \\ a + 3b + 4c \leq 150 \\ 2a + 2b + 2c \leq 140 \\ \qquad\quad c \geq 0,\ b \geq 0,\ c \geq 0 \end{cases}$$

The dual problem is to minimize $120u + 150v + 140w$ subject to the constraints

$$\begin{cases} 2u + v + 2w \geq 8 \\ u + 3v + 2w \geq 10 \\ u + 4v + 2w \geq p \\ \qquad\quad u \geq 0,\ v \geq 0,\ w \geq 0 \end{cases}$$

The original solution, with $u = 0$, $v = 1$, $w = \dfrac{7}{2}$, will be optimal if

$$0 + 4(1) + 2\left(\frac{7}{2}\right) \geq p. \text{ Since}$$

$$0 + 4(1) + 2\left(\frac{7}{2}\right) = \$11, \text{ that is the profit}$$

per tennis racket that needs to be realized to justify the diversification.

18. The new problem is to maximize $70a + 210b + 140c + pd$, where d is the number of Brand D stereo systems and p is the profit per system, subject to the

$$\text{constraints} \begin{cases} a + b + c + d \leq 100 \\ 5a + 4b + 4c + 3d \leq 480 \\ 40a + 20b + 30c + 30d \leq 3200 \\ a \geq 0,\ b \geq 0,\ c \geq 0,\ d \geq 0 \end{cases}.$$

The dual problem is to maximize $100u + 480v + 3200w$, subject to the

$$\text{constraints} \begin{cases} u + 5v + 40w \geq 70 \\ u + 4v + 20w \geq 210 \\ u + 4v + 30w \geq 140 \\ u + 3v + 30w \geq p \\ u \geq 0,\ v \geq 0,\ w \geq 0 \end{cases}.$$

The original solution, with $u = 210$, $v = 0$, $w = 0$, will still be optimal if $210 + 3(0) + 30(0) \geq p$. Since $210 + 3(0) + 30(0) = \$210$, that is the required profit per brand D system. The original solution was to sell 100 units of brand B at a profit of \$210 each. Brand D units have to be at least this profitable. (Differences in storage space and commission turned out not to matter.)

Chapter 5

Exercises 1

1. **a.** $S' = \{5, 6, 7\}$

   b. $S \cup T = \{1, 2, 3, 4, 5, 7\}$

   c. $S \cap T = \{1, 3\}$

   d. $S' \cap T = \{5, 7\}$

3. **a.** $R \cup S = \{a, b, c, d, e, f\}$

   b. $R \cap S = \{c\}$

   c. $S \cap T = \varnothing$

5. $\varnothing, \{1\}, \{2\}, \{1, 2\}$

7. **a.** $M \cap F = \{$all male college students who like football$\}$

   b. $M' = \{$all female college students$\}$

   c. $M' \cap F' = \{$all female college students who don't like football$\}$

   d. $M \cup F = \{$all male college students or all college students who like football$\}$

9. **a.** $S = \{1950, 1951, 1958, 1963, 1967, 1975, 1976\}$

   b. $T = \{1950, 1951, 1954, 1955, 1958, 1961, 1963, 1967, 1975, 1976\}$

   c. $S \cap T = \{1950, 1951, 1958, 1963, 1967, 1975, 1976\}$

   d. $S' \cap T = \{1954, 1955, 1961\}$

   e. $S \cap T' = \varnothing$

11. From 1950 to 1977, whenever the Standard and Poor's Index increased by 2% or more during the first five days, it also increased by 16% or more for that year.

13. **a.** $R \cup S = \{a, b, c, e\}$
    $(R \cup S)' = \{d, f\}$

    b. $R \cup S \cup T = \{a, b, c, e, f\}$

    c. $R \cap S = \{a, c\}$
    $R \cap S \cap T = (R \cap S) \cap T = \varnothing$

    d. $T' = \{a, b, c, d\}$
          $R \cap S \cap T' = (R \cap S) \cap T' = \{a, c\}$

    e. $R' = \{d, e, f\};\ S \cap T = \{e\}$
          $R' \cap S \cap T = R' \cap (S \cap T) = \{e\}$

    f. $S \cup T = \{a, c, e, f\}$

    g. $R \cup S = \{a, b, c, e\};$
          $R \cup T = \{a, b, c, e, f\}$
          $(R \cup S) \cap (R \cup T) = \{a, b, c, e\}$

    h. $R \cap S = \{a, c\};\ R \cap T = \varnothing$
          $(R \cap S) \cup (R \cap T) = \{a, c\}$

    i. $R' = \{d, e, f\};\ T' = \{a, b, c, d\}$
          $R' \cap T' = \{d\}$

15. $(S')' = S$

17. $S \cup S' = U$

19. $T \cap S \cap T' = S \cap (T \cap T') = S \cap \varnothing = \varnothing$

21. {divisions that had increases in labor costs or total revenue} $= L \cup T$

23. {divisions that made a profit despite an increase in labor costs} $= L \cap P$

25. {profitable divisions with increases in labor costs and total revenue} $= P \cap L \cap T$

27. {applicants who have not received speeding tickets} $= S'$

29. {applicants who have received speeding tickets, caused accidents, or been arrested for drunk driving}
    $= S \cup A \cup D$

31. {applicants who have not both caused accidents and received speeding tickets but who have been arrested for drunk driving}
    $= (A \cap S)' \cap D$

33. $A \cap D = $ {male students at Mount College}

35. $A \cap B = $ {people who are both teachers and students at Mount College}

37. $A \cup C' = A \cup D = $ {males or students at Mount College}

39. $D' = C = $ {females at Mount College}

41. {people who don't like strawberry ice cream} $= S'$

43. {people who like vanilla or chocolate but not strawberry ice cream} $= (V \cup C) \cap S'$

45. {people who like neither chocolate nor vanilla ice cream} $= (V \cup C)'$

47. a. $R = \{B, C, D, E\}$

    b. $S = \{C, D, E, F\}$

    c. $T = \{A, D, E, F\}$

    d. $R' = \{A, F\}$
       $R' \cup S = \{A, C, D, E, F\}$

    e. $R' \cap T = \{A, F\}$

    f. $R \cap S = \{C, D, E\}$
       $R \cap S \cap T = (R \cap S) \cap T = \{D, E\}$

49. Any subset with 2 as an element is an example. Possible answer: $\{2\}$

51. If S is a subset of T, then $S \cup T = T$.

Exercises 2

  1. $n(S \cup T) = n(S) + n(T) - n(S \cap T)$
     $= 5 + 4 - 2 = 7$

  3. $n(S \cup T) = n(S) + n(T) - n(S \cap T)$
     $15 = 7 + 8 - n(S \cap T)$
     $n(S \cap T) = 7 + 8 - 15 = 0$

  5. $n(S \cup T) = n(S) + n(T) - n(S \cap T)$
     $13 = n(S) + 7 - 5$
     $n(S) = 13 - 7 + 5 = 11$

  7. S is a subset of T.

  9. Let $P = \{$adults in South America fluent in Portuguese$\}$ and
     $S = \{$adults in South America fluent in Spanish$\}$.
     Then $P \cup S = \{$adults in South America fluent in Portuguese or Spanish$\}$ and
     $P \cap S = \{$adults in South America fluent in Portuguese and Spanish$\}$.
     $n(P) = 99, n(S) = 95, \ n(P \cup S) = 180$ (numbers in millions)
     $n(P \cup S) = n(P) + n(S) - n(P \cap S)$
     $180 = 99 + 95 - n(P \cap S)$
     $n(P \cap S) = 99 + 95 - 180 = 14$
     14 million are fluent in both languages.

11. Let $M = \{$members of the MAA$\}$ and
$A = \{$members of the AMS$\}$.
Then
$M \cup A = \{$members of the MAA or AMS$\}$ and
$M \cap A = \{$members of both the MAA and AMS$\}$.
$n(A) = 29,000, \; n(M \cap A) = 7000, \; n(M \cup A) = 52,000$
$n(M \cup A) = n(M) + n(A) - n(M \cap A)$
$52,000 = n(M) + 29,000 - 7000$
$n(M) = 52,000 - 29,000 + 7000 = 30,000$
30,000 people belong to the MAA.

13. Let $A = \{$cars with automatic transmission$\}$ and $P = \{$cars with power steering$\}$.
Then $A \cup P = \{$cars with automatic transmission or power steering$\}$ and
$A \cap P = \{$cars with both automatic transmission and power steering$\}$,
$n(A) = 325, n(P) = 216, n(A \cap P) = 89$
$n(A \cup P) = n(A) + n(P) - n(A \cap P)$
$= 325 + 216 - 89 = 452$
452 cars were manufactured with at least one of the two options.

15. Consists of points not in S and not in T.

17. Consists of points in S or not in T.

19. $(S \cap T)' = S' \cup T'$
Consists of points not in S or not in T.

21. Consists of points in S but not in T or points in T but not in S.

23. $S \cup (S \cap T) = S$
Consists of points in S.

25. $S \cup S' = U$
Consists of all points.

27. Consists of points in R and S but not in T.

29. Consists of points in R or points in both S and T.

31. Consists of points in R but not in S or points in both R and T.

33. Consists of points in both R and T.

35. Consists of points not in R, S, and T.

37. Consists of points in R and T or points in S but not T.

39. $S' \cup (S \cap T)' = S' \cup S' \cup T' = S' \cup T'$

41. $(S' \cup T)' = S \cap T'$

43. $T \cup (S \cap T)' = T \cup S' \cup T'$
$= (T \cup T') \cup S' = U$

45. S'

47. $R \cap T$

49. $R' \cap S \cap T$

51.

$T \cup (R \cap S')$

53. First draw a Venn diagram for $(R \cap S') \cup (S \cap T') \cup (T \cap R')$.

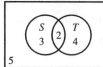

The set consists of the complement.
$(R \cap S \cap T) \cup (R' \cap S' \cap T')$

55. People who are not illegal aliens or everyone over the age of 18 who is employed.

57. Everyone over the age of 18 who is unemployed

59. B is a subset of A', so $A' \cup B = A'$.
Noncitizens who are unemployed

Exercises 3

1. $n(S \cap T') = n(S) - n(S \cap T) = 5 - 2 = 3$
$n(S' \cap T) = n(T) - n(S \cap T) = 6 - 2 = 4$
$n(S \cup T) = n(S) + n(T) - n(S \cap T)$
$= 5 + 6 - 2 = 9$
$n(S' \cap T') = n(U) - n(S \cup T) = 14 - 9 = 5$

3. $n(S \cap T) = n(S) + n(T) - n(S \cup T) = 12 + 14 - 18 = 8$
$n(S \cap T') = n(S) - n(S \cap T) = 12 - 8 = 4$
$n(S' \cap T) = n(T) - n(S \cap T) = 14 - 8 = 6$
$n(S' \cap T') = n(U) - n(S \cup T) = 20 - 18 = 2$

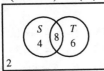

5. $n(S \cup T) = n(U) - n(S' \cap T') = 75 - 40 = 35$
$n(S \cap T) = n(S) + n(T) - n(S \cup T) = 15 + 25 - 35 = 5$
$n(S \cap T') = n(S) - n(S \cap T) = 15 - 5 = 10$
$n(S' \cap T) = n(T) - n(S \cap T) = 25 - 5 = 20$

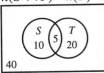

7. $n(S \cap T) = n(S) + n(T) - n(S \cup T) = 3 + 4 - 6 = 1$
$n(U) = n(S' \cup T') + n(S \cap T) = 9 + 1 = 10$
$n(S \cap T') = n(S) - n(S \cap T) = 3 - 1 = 2$
$n(S' \cap T) = n(T) - n(S \cap T) = 4 - 1 = 3$
$n(S' \cap T') = n(U) - n(S \cup T) = 10 - 6 = 4$

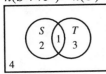

9. $n(R \cap S \cap T') = n(R \cap S) - n(R \cap S \cap T) = 7 - 2 = 5$
$n(R \cap S' \cap T) = n(R \cap T) - n(R \cap S \cap T) = 6 - 2 = 4$
$n(R' \cap S \cap T) = n(S \cap T) - n(R \cap S \cap T) = 5 - 2 = 3$
$n(R \cap S' \cap T') = n(R) - n(R \cap S \cap T') - n(R \cap S' \cap T) - n(R \cap S \cap T) = 17 - 5 - 4 - 2 = 6$
$n(R' \cap S \cap T') = n(S) - n(R \cap S \cap T') - n(R' \cap S \cap T) - n(R \cap S \cap T) = 17 - 5 - 3 - 2 = 7$
$n(R' \cap S' \cap T) = n(T) - n(R \cap S' \cap T) - n(R' \cap S \cap T) - n(R \cap S \cap T) = 17 - 4 - 3 - 2 = 8$
$n(R \cup S \cup T) = 6 + 7 + 8 + 5 + 4 + 3 + 2 = 35$
$n(R' \cap S' \cap T') = n(U) - n(R \cup S \cup T) = 44 - 35 = 9$

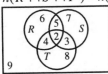

11. $n(R) = n(R \cup S) + n(R \cap S) - n(S) = 21 + 7 - 14 = 14$
     $n(U) = n(R) + n(R') = 22 + 14 = 36$
     $n(R \cap S \cap T') = n(R \cap S) - n(R \cap S \cap T) = 7 - 5 = 2$
     $n(R \cap S' \cap T) = n(R \cap T) - n(R \cap S \cap T) = 11 - 5 = 6$
     $n(R' \cap S \cap T) = n(S \cap T) - n(R \cap S \cap T) = 9 - 5 = 4$
     $n(R \cap S' \cap T') = n(R) - n(R \cap S \cap T') - n(R \cap S' \cap T) - n(R \cap S \cap T) = 14 - 2 - 6 - 5 = 1$
     $n(R' \cap S \cap T') = n(S) - n(R \cap S \cap T') - n(R' \cap S \cap T) - n(R \cap S \cap T) = 14 - 2 - 4 - 5 = 3$
     $n(R' \cap S' \cap T) = n(T) - n(R \cap S' \cap T) - n(R' \cap S \cap T) - n(R \cap S \cap T) = 22 - 6 - 4 - 5 = 7$
     $n(R \cup S \cup T) = 1 + 3 + 7 + 2 + 6 + 4 + 5 = 28$
     $n(R' \cap S' \cap T') = n(U) - n(R \cup S \cup T) = 36 - 28 = 8$

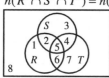

13. Let $U = \{$high school students surveyed$\}$, $F = \{$students who like folk music$\}$,
     and $C = \{$students who like classical music$\}$.
     $n(U) = 70;\ n(F) = 35;\ n(C) = 15;\ n(F \cap C) = 5$
     $n(F \cup C) = n(F) + n(C) - n(F \cap C) = 35 + 15 - 5 = 45$
     $n((F \cup C)') = n(U) - n(F \cup C) = 70 - 45 = 25$
     25 students do not like either folk or classical music.

15. Let $U = \{$students in finite math$\}$, $M = \{$male students$\}$, $B = \{$students who are business majors$\}$,
     and $F = \{$first-year students$\}$.
     $n(U) = 35;\ n(M) = 22;\ n(B) = 19;\ n(F) = 27;$
     $n(M \cap B) = 14;\ n(M \cap F) = 17;\ n(B \cap F) = 15;\ n(M \cap B \cap F) = 11$
     Draw a Venn diagram as shown.

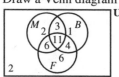

     $n((M \cup B \cup F)') = 2$
     There are two upperclass women nonbusiness majors.
     $n(M' \cap B) = 1 + 4 = 5$
     There are five women business majors.

17. Let $U = \{$college students surveyed$\}$, $F = \{$first-year students$\}$, $D = \{$voted Democratic$\}$,
     $n(U) = 100;\ n(F) = 50;\ n(D) = 55$
     $n(F' \cap D') = n((F \cup D)') = 25$
     $n(F \cup D) = n(U) - n((F \cup D)') = 100 - 25 = 75$
     $n(F \cap D) = n(F) + n(D) - n(F \cup D) = 50 + 55 - 75 = 30$
     30 freshman voted Democratic.

19. Let $U = \{\text{students}\}$, $D = \{\text{students who passed the diagnostic test}\}$,
$C = \{\text{students who passed the course}\}$.
$n(U) = 30;\ n(D) = 21;\ n(C) = 23;\ n(D \cap C') = 2$
$n(D \cap C) = n(D) - n(D \cap C') = 21 - 2 = 19$
$n(D' \cap C) = n(C) - n(D \cap C) = 23 - 19 = 4$
Four students passed the course even though they failed the diagnostic test.

21. Let $U = \{\text{lines}\}$, $V = \{\text{lines with verbs}\}$, $A = \{\text{lines with adjectives}\}$
$n(U) = 14;\ n(V) = 11;\ n(A) = 9;$
$n(V \cap A) = 7$
$n(V \cap A') = n(V) - n(V \cap A) = 11 - 7 = 4$
Four lines have a verb but no adjective.
$n(V' \cap A) = n(A) - n(V \cap A) = 9 - 7 = 2$
Two lines have an adjective but no verb.
$n(V \cup A) = n(V) + n(A) - n(V \cap A) = 11 + 9 - 7 = 13$
$n((V \cup A)') = n(U) - n(V \cap A) = 14 - 13 = 1$
One line has neither an adjective nor a verb.

For Exercises 23–25, let $U = \{\text{students who took the math exam}\}$,
$F = \{\text{students who correctly answered the first question}\}$,
$S = \{\text{students who correctly answered the second question}\}$. Then $n(U) = 130$, $n(F) = 90$, $n(S) = 62$,
$n(F \cap S) = 50$. Draw and complete the Venn diagram as follows.

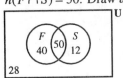

23. $n((F \cup S)') = 28$

25. $n(S \cap F') = 12$

For Exercises 27–31, let $U = \{\text{football cards}\}$, $N = \{\text{players in the NFL}\}$, $D = \{\text{players who played defense}\}$.
Then $n(U) = 2200$, $n(N) = 1500$, $n(D) = 900$, $n(N \cap D) = 400$. Draw and complete the Venn diagram as
follows.

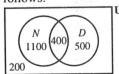

27. $n(N \cup D) = 1100 + 500 + 400 = 2000$

29. $n(N' \cap D') = n((N \cup D)') = 200$

31. $n((N \cap D') \cup (N' \cap D)) = 1100 + 500 = 1600$

For Exercises 33–39, let U = {surveyed students}, R = {students who like rock},
C = {students who like country}, L = {students who like classical music}. Then $n(U) = 190, n(R) = 114$,
$n(C) = 50, n(L) = 41, n(R \cap C) = 14, n(R \cap L) = 15, n(C \cap L) = 11, n(R \cap C \cap L) = 5$. Draw and complete
the Venn diagram as follows.

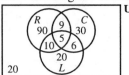

33. $n(R \cap C' \cap L') = 90$

35. $n(R' \cap C \cap L) = 6$

37. $n((R \cap C' \cap L') \cup (R' \cap C \cap L') \cup (R' \cap C' \cap L)) = 90 + 30 + 20 = 140$

39. $n((R \cap C) \cup (R \cap L) \cup (C \cap L)) = 9 + 10 + 6 + 5 = 30$

For Exercises 41–45, let U = {people surveyed}, R = {people who learned from radio},
T = {people who learned from television}, N = {people who learned from newspapers}. Then $n(U) = 400$,
$n(R) = 180, n(T) = 190, n(N) = 190, n(R \cap T) = 80, n(R \cap N) = 90, n(T \cap N) = 50, n(R \cap T \cap N) = 30$.
Draw and complete the Venn diagram as follows.

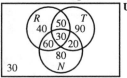

41. $n((R \cap N') \cup (R' \cap N)) = 40 + 50 + 80 + 20 = 190$

43. $n((R \cup T) \cap N') = 40 + 90 + 50 = 180$

45. $n((R \cap T' \cap N') \cup (R' \cap T \cap N') \cup (R' \cap T' \cap N)) = 40 + 90 + 80 = 210$

47. Let U = {executives}, F = {executives who read *Fortune*}, T = {executives who read *Time*},
M = {executives who read *Money*}.
$n(U) = 180, n(F) = 75, n(T) = 70, n(M) = 55, n(F \cap T) = 25, n(T \cap M) = 25, n(R \cap T \cap N) = 5$
Draw and complete the Venn diagram as follows.

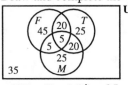

$n((F \cup T \cup M)') = 35$

For Exercises 49–53, $n(U) = 4000$, $n(F) = 2000$, $n(S) = 3000$, $n(L) = 500$, $n(F \cap S) = 1500$, $n(F \cap L) = 300$, $n(S \cap L) = 200$, $n(F \cap S \cap L) = 50$. Draw and complete the Venn diagram as follows.

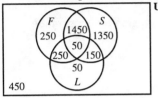

49. $L \cap (F \cup S) = 250 + 150 + 50 = 450$

51. $L \cup S \cup F' = 4000 - 250 = 3750$

53. $F \cap S' \cap L' = 250$

Exercises 4

1. $3 \cdot 5 = 15$ routes

3. $26 \cdot 26 = 676$ words

5. $20 \cdot 19 = 380$ tickets

7. $5 \cdot 4 = 20$ ways

9. $5 \cdot 4 \cdot 3 \cdot 2 \cdot 1 = 120$ ways

11. $2 \cdot 2 \cdot 2 \cdot 2 \cdot 2 \cdot 2 = 2^6 = 64$ sequences

13. $20 \cdot 19 \cdot 18 = 6840$ possibilities

15. $30 \cdot 29 = 870$ ways

17. $4 \cdot 3 \cdot 2 \cdot 1 = 24$ words

19. $2 \cdot 2 \cdot 2 \cdot 2 \cdot 2 = 2^5 = 32$ ways

21. $3 \cdot 12 \cdot 10 \cdot 10 \cdot 10 \cdot 10 = 360{,}000$ serial numbers

23. a. $7 \cdot 7 \cdot 7 \cdot 7 = 2401$ words

    b. $7 \cdot 6 \cdot 5 \cdot 4 = 840$ words

    c. $1 \cdot 7 \cdot 7 \cdot 7 = 343$ words

    d. Let operation 1 be choosing the letter at the end.
    $2 \cdot 6 \cdot 5 \cdot 4 = 240$ words

25. a. $9 \cdot 8 \cdot 7 \cdot 6 \cdot 5 \cdot 4 \cdot 3 \cdot 2 \cdot 1 = 362{,}880$ batting orders

b. $8 \cdot 7 \cdot 6 \cdot 5 \cdot 4 \cdot 3 \cdot 2 \cdot 1 \cdot 1 = 40{,}320$ batting orders

c. $1 \cdot 6 \cdot 5 \cdot 4 \cdot 3 \cdot 2 \cdot 1 \cdot 1 \cdot 1 = 720$ batting orders

27. A Venn diagram with three circles has 8 regions that can be shaded. Let each region be an operation. Each operation can be performed two ways, shaded or not shaded.
$2 \cdot 2 \cdot 2 \cdot 2 \cdot 2 \cdot 2 \cdot 2 \cdot 2 = 256$ ways

29. Each statement has three options.
$3 \cdot 3 \cdot 3 \cdot 3 \cdot 3 \cdot 3 = 3^6 = 729$ ways

31. $6 \cdot 7 \cdot 4 = 168$ days

33. $3 \cdot 6 = 18$

35. The first house has four choices of color. Each house after that has only 3 choices.
$4 \cdot 3 \cdot 3 \cdot 3 \cdot 3 \cdot 3 = 972$ ways

37. $4 \cdot 4 \cdot 4 \cdot 4 \cdot 4 \cdot 4 \cdot 4 \cdot 4 \cdot 4 \cdot 4 = 4^{10} = 1{,}048{,}576$ ways

39. $4 \cdot 4 = 16$ ways

41. Each bottle has five possible locations to be delivered.
$5 \cdot 5 \cdot 5 \cdot 5 \cdot 5 \cdot 5 \cdot 5 \cdot 5 \cdot 5 \cdot 5 = 5^{10} = 9{,}765{,}625$ ways

43. If every voter votes on each of the items, there are $7 \cdot 4 \cdot 12 = 336$ ballots.
If voters can choose to leave an item blank, there are $8 \cdot 5 \cdot 13 = 520$ ballots.

45. $5 \cdot 5 \cdot 2 = 50$ bags

47. $5 \cdot 2 \cdot 3 \cdot 2 \cdot 2 \cdot 8 = 960$ cars

49. James has 100 choices. Janet has 14 choices. Sandy has 4 choices. There are $100 \cdot 14 \cdot 4 = 5600$ triples.

51. A basket can have 0, 1, 2, 3, or 4 apples, so there are 5 possibilities for apples. Likewise, there are 4 possibilities for pears and 5 possibilities for oranges. Including the possibility of no fruit in a basket, there are $5 \cdot 4 \cdot 5 = 100$ baskets. Thus there are $100 - 1 = 99$ baskets if the basket must contain at least one piece of fruit.

53. Let each operation be choosing a position for a player.
Let operation 1 be choosing a position for the child who can only be catcher. There is only 1 choice.
Let operation 2 be choosing a position for one of the two children who can pitch or play outfield. There are 2 choices.
Let operation 3 be choosing a position for the other child who can pitch or play outfield. There is only one choice left.
Let operation 4 be choosing a position for one of the three children who can play only as infielders. There are 4 choices.
Let operation 5 be choosing a position for one of the remaining two children who can play only as infielders. There are 3 choices.
Let operation 6 be choosing a position for the remaining child who can play only as an infielder. There are 2 choices.

Let operation 7 be choosing a position for one of the remaining children. There are 3 choices remaining.
Let operation 8 be choosing a position for one of the remaining children. There are 2 choices remaining.
Let operation 9 be choosing a position for the remaining child. There is only 1 choice remaining.
Thus, there are $1 \cdot 2 \cdot 1 \cdot 4 \cdot 3 \cdot 2 \cdot 3 \cdot 2 \cdot 1 = 288$ lineups.

55. Let n be the number of each article of clothing. The possible number of combinations is $n \cdot n \cdot n = n^3$. We want $n^3 > 365$. This occurs when $n > 7$. Thus she needs 8 of each article of clothing.

57. If the number is of the form $4xxx$, there are 4 possible even numbers for the last digit. Then there are $4 \cdot 8 \cdot 7 = 224$ such numbers. Likewise for $6xxx$. If the number is of the form $5xxx$, there are 5 possible even numbers for the last digit. There are $5 \cdot 8 \cdot 7 = 280$ such numbers. Thus there are $224 + 224 + 280 = 728$ integers.

Exercises 5

1. $P(4, 2) = 4 \cdot 3 = 12$

3. $P(6, 3) = 6 \cdot 5 \cdot 4 = 120$

5. $C(10, 3) = \dfrac{P(10, 3)}{3!} = \dfrac{10 \cdot 9 \cdot 8}{3 \cdot 2 \cdot 1} = 120$

7. $C(5, 4) = \dfrac{P(5, 4)}{4!} = \dfrac{5 \cdot 4 \cdot 3 \cdot 2}{4 \cdot 3 \cdot 2 \cdot 1} = 5$

9. $P(5, 1) = 5$

11. $P(n, 1) = n$

13. $C(4, 4) = \dfrac{P(4, 4)}{4!} = \dfrac{4 \cdot 3 \cdot 2 \cdot 1}{4 \cdot 3 \cdot 2 \cdot 1} = 1$

15. $C(n, n-2) = \dfrac{P(n, n-2)}{(n-2)!}$

$= \dfrac{n \cdot (n-1) \cdots 4 \cdot 3}{(n-2) \cdot (n-3) \cdots 2 \cdot 1} = \dfrac{n \cdot (n-1)}{2 \cdot 1}$

$= \dfrac{n^2 - n}{2}$

17. $6! = 6 \cdot 5 \cdot 4 \cdot 3 \cdot 2 \cdot 1 = 720$

19. $\dfrac{9!}{7!} = \dfrac{9 \cdot 8 \cdot 7 \cdot 6 \cdot 5 \cdot 4 \cdot 3 \cdot 2 \cdot 1}{7 \cdot 6 \cdot 5 \cdot 4 \cdot 3 \cdot 2 \cdot 1} = 9 \cdot 8 = 72$

21. $P(4, 4) = 4 \cdot 3 \cdot 2 \cdot 1 = 24$ ways

23. $C(9, 2) = \dfrac{P(9, 2)}{2!} = \dfrac{9 \cdot 8}{2 \cdot 1} = 36$ selections

25. $P(15, 3) = 15 \cdot 14 \cdot 13 = 2730$ ways

27. $C(10, 4) = \dfrac{P(10, 4)}{4!} = \dfrac{10 \cdot 9 \cdot 8 \cdot 7}{4 \cdot 3 \cdot 2 \cdot 1}$
$= 210$ deluxe chocolate banana splits

29. $C(10, 6) = \dfrac{P(10, 6)}{6!} = \dfrac{10 \cdot 9 \cdot 8 \cdot 7 \cdot 6 \cdot 5}{6 \cdot 5 \cdot 4 \cdot 3 \cdot 2 \cdot 1}$
$= 210$ ways

31. $P(5, 5) = 5 \cdot 4 \cdot 3 \cdot 2 \cdot 1 = 120$ ways

33. $C(8, 2) = \dfrac{P(8, 2)}{2!} = \dfrac{8 \cdot 7}{2 \cdot 1} = 28$ games

35. $P(10, 5) = 10 \cdot 9 \cdot 8 \cdot 7 \cdot 6 = 30{,}240$ ways

37. $C(10, 5) = \dfrac{P(10, 5)}{5!} = \dfrac{10 \cdot 9 \cdot 8 \cdot 7 \cdot 6}{5 \cdot 4 \cdot 3 \cdot 2 \cdot 1}$
$= 252$ ways

39. $P(35, 5) = 35 \cdot 34 \cdot 33 \cdot 32 \cdot 31$
$= 38{,}955{,}840$ ways

41. $C(100, 3) = \dfrac{P(100, 3)}{3!} = \dfrac{100 \cdot 99 \cdot 98}{3 \cdot 2 \cdot 1}$
$= 161{,}700$ samples are possible
$C(7, 3) = \dfrac{P(7, 3)}{3!} = \dfrac{7 \cdot 6 \cdot 5}{3 \cdot 2 \cdot 1}$
$= 35$ samples consist of all defective diskettes

43. $P(26, 3) = 26 \cdot 25 \cdot 24 = 15,600$ words

45. $C(100, 5) = \dfrac{P(100, 5)}{5!}$

$= \dfrac{100 \cdot 99 \cdot 98 \cdot 97 \cdot 96}{5 \cdot 4 \cdot 3 \cdot 2 \cdot 1}$

$= 75,287,520$ ways

47. $C(52, 5) = \dfrac{P(52, 5)}{5!} = \dfrac{52 \cdot 51 \cdot 50 \cdot 49 \cdot 48}{5 \cdot 4 \cdot 3 \cdot 2 \cdot 1}$

$= 2,598,960$ hands

49. There are 13 clubs in a deck.

$C(13, 5) = \dfrac{P(13, 5)}{5!} = \dfrac{13 \cdot 12 \cdot 11 \cdot 10 \cdot 9}{5 \cdot 4 \cdot 3 \cdot 2 \cdot 1}$

$= 1287$ hands

51. $C(20, 3) = \dfrac{P(20, 3)}{3!} = \dfrac{20 \cdot 19 \cdot 18}{3 \cdot 2 \cdot 1}$

$= 1140$ ways

53. $P(5, 5) = 5 \cdot 4 \cdot 3 \cdot 2 \cdot 1 = 120$ ways

55. $C(8, 4) = \dfrac{P(8, 4)}{4!} = \dfrac{8 \cdot 7 \cdot 6 \cdot 5}{4 \cdot 3 \cdot 2 \cdot 1} = 70$ ways

57. Moe has

$C(9, 2) = \dfrac{P(9, 2)}{2!} = \dfrac{9 \cdot 8}{2 \cdot 1} = 36$ choices.

Joe has

$C(7, 3) = \dfrac{P(7, 3)}{3!} = \dfrac{7 \cdot 6 \cdot 5}{3 \cdot 2 \cdot 1} = 35$ choices.

Thus Joe is correct.

59. This is the same as ordering 12 different pictures. Then there are
$P(12, 12) = 12!$
$= 479,001,600$ arrangements.

61. There is only one possibility for the pitcher and there are two possibilities for the first baseman. Of the remaining seven players, there are $P(7, 7) = 7! = 5040$ possibilities. Thus there are $1 \cdot 2 \cdot 5040 = 10,080$ batting orders.

63. $P(5, 5) = 5! = 120$ ways

65. Of cards from two suits, there are

$C(26, 5) = \dfrac{P(26, 5)}{5!} = \dfrac{26 \cdot 25 \cdot 24 \cdot 23 \cdot 22}{5 \cdot 4 \cdot 3 \cdot 2 \cdot 1}$

$= 65,780$ five-card combinations. Of those,

$C(13, 5) = \dfrac{P(13, 5)}{5!} = \dfrac{13 \cdot 12 \cdot 11 \cdot 10 \cdot 9}{5 \cdot 4 \cdot 3 \cdot 2 \cdot 1}$

$= 1287$ consist of cards from one of the two suits. Thus there are
$65,780 - 2(1287) = 63,206$ five-card combinations with exactly two suits, given the two suits.
Since there are

$C(4, 2) = \dfrac{P(4, 2)}{2!} = \dfrac{4 \cdot 3}{2 \cdot 1} = 6$ possible

two-suit combinations, there are
$6 \cdot 63,206 = 379,236$ five-card combinations from exactly two suits.

67. The product is even if at least one of the digits is even. There are $10^5 = 100,000$ possible ZIP codes altogether. There are $5^5 = 3125$ possible ZIP codes consisting of all odd digits. Thus there are $100,000 - 3125 = 96,875$ ZIP codes in which the product of the digits is even.

69. Number of programs for a student on the semester system:
$C(752, 5)C(747, 5)C(742, 5)C(737, 5)C(732, 5)C(727, 5)C(722, 5)C(717, 5)$
Number of programs for a student on the trimester system:
$C(937, 3)C(934, 3)C(931, 3)C(928, 3)C(925, 3)C(922, 3)C(919, 3)C(916, 3)C(913, 3)C(910, 3)C(907, 3)$
$C(904, 3)$
Using a computer to compare the two quantities, semester system students have a greater number of different programs.

71. a. $C(45, 5) = 1,221,759$ tickets

  b. $C(100, 4) = 3,921,225$ tickets

  c. First lottery

  d. Number of possible tickets for the first lottery:
  $P(45, 5) = 146,611,080$
  Number of possible tickets for the second lottery:
  $P(100, 4) = 94,109,400$
  Thus the second lottery is better if order matters.

Exercises 6

1. a. $2^6 = 64$ outcomes

  b. $C(6, 3) = \dfrac{6 \cdot 5 \cdot 4}{3 \cdot 2 \cdot 1} = 20$ outcomes

  c. These are the outcomes with exactly no tails, one tail, or two tails.
  $$C(6, 0) + C(6, 1) + C(6, 2) = 1 + \frac{6}{1} + \frac{6 \cdot 5}{2 \cdot 1} = 22 \text{ outcomes}$$

  d. These are all of the outcomes minus the outcomes with exactly no heads or one head.
  $$2^6 - C(6, 0) - C(6, 1) = 64 - 1 - \frac{6}{1} = 57 \text{ outcomes}$$

3. Four of the nine blocks in the route must be "south."
  $$C(9, 4) = \frac{9 \cdot 8 \cdot 7 \cdot 6}{4 \cdot 3 \cdot 2 \cdot 1} = 126 \text{ routes}$$

5. a. $C(10, 3) = \dfrac{10 \cdot 9 \cdot 8}{3 \cdot 2 \cdot 1} = 120$ samples

  b. $C(8, 3) = \dfrac{8 \cdot 7 \cdot 6}{3 \cdot 2 \cdot 1} = 56$ samples

  c. From part (b) there are 56 samples with no rotten apples.
  Samples that contain at least one rotten apple:
  $120 - 56 = 64$ samples

7. Consider this as a sequence of three operations.
  Assign 25 students to dorm A: $C(100, 25)$
  Assign 40 of the remaining 75 students to dorm B: $C(75, 40)$
  Assign 35 of the remaining 35 students to dorm C: $C(35, 35) = 1$
  Ways to assign all 100 students: $C(100, 25) \cdot C(75, 40)$

9. Two of the five blocks from A to C must be "south."
Routes from A to C:
$$C(5, 2) = \frac{5 \cdot 4}{2 \cdot 1} = 10 \text{ routes}$$
Two of the four blocks from C to B must be "south."
Routes from C to B:
$$C(4, 2) = \frac{4 \cdot 3}{2 \cdot 1} = 6 \text{ routes}$$
Routes from A to B passing through C:
$10 \cdot 6 = 60$ routes

11. First select the states from which each senator comes from. There are
$$C(50, 5) = \frac{50 \cdot 49 \cdot 48 \cdot 47 \cdot 46}{5 \cdot 4 \cdot 3 \cdot 2 \cdot 1} = 2,118,760$$
ways to do this.
Once the states have been selected, each state has two possible choices of senators, so there are $2^5 = 32$ possible committees once the states have been chosen. Ways a committee of five senators can be selected so that no two members are from the same state:
$2,118,760 \cdot 32 = 67,800,320$ ways

13. $C(4, 3) \cdot C(6, 3) = \dfrac{4 \cdot 3 \cdot 2}{3 \cdot 2 \cdot 1} \cdot \dfrac{6 \cdot 5 \cdot 4}{3 \cdot 2 \cdot 1}$
$= 4 \cdot 20 = 80$ ways

15. $C(20, 12) \cdot C(8, 2) = 125,970 \cdot 28$
$= 3,527,160$ ways

17. $C(9, 5) \cdot C(8, 6) = 126 \cdot 28 = 3528$ ways

19. $C(15, 3) \cdot C(12, 8) \cdot C(4, 4) = 455 \cdot 495 \cdot 1$
$= 225,225$ ways

21. $P(26, 6) = 26 \cdot 25 \cdot 24 \cdot 23 \cdot 22 \cdot 21$
$= 165,765,600$ plates

23. $P(7, 3) = 7 \cdot 6 \cdot 5 = 210$ words

25. $P(26, 3) \cdot P(10, 3) = 15,600 \cdot 720$
$= 11,232,000$ plates

27. First select the lineup of the four classes. There are $4! = 24$ ways to do this. Then arrange three of the four freshmen. There are $P(4, 3) = 24$ ways to do this. Then arrange three of the five sophomores. There are $P(5, 3) = 60$ ways to do this. Then arrange three of the six juniors. There are $P(6, 3) = 120$ ways to do this. Then arrange three of the seven seniors. There are $P(7, 3) = 210$ ways to do this.
Number of different pictures:
$24 \cdot 24 \cdot 60 \cdot 120 \cdot 210 = 870,912,000$ pictures

29. First select the order of the towns. There are $2! = 2$ ways to do this. Then select the order in the larger town. There are $3! = 6$ ways to do this. Then select the order in the smaller town. There are $2! = 2$ ways to do this.
Ways to schedule his stops:
$2 \cdot 6 \cdot 2 = 24$ ways

31. $C(4, 3) \cdot C(4, 2) = 4 \cdot 6 = 24$ hands

33. First select the first denomination. There 13 ways to do this. Then select three of the denomination. There are $C(4, 3) = 4$ ways to do this. Then select the second denomination. There are 12 ways to do this. Then select two of the denomination. There are $C(4, 2) = 6$ ways to do this.
Number of poker hands:
$13 \cdot 4 \cdot 12 \cdot 6 = 3744$ hands

35. The digits that can be used in a detour-prone ZIP code are 0, 1, 6, 8, and 9. There are $5^5 = 3125$ ZIP codes that use only those digits. Any ZIP code of the form XYZYX where X, Y, and Z are chosen from 0, 1, and 8 is not detour-prone. There are $3^3 = 27$ such ZIP codes. If X represents the digit 6 or 9 then X′ represents the digit 9 or 6, respectively. Any ZIP code of the form XYZYX′ where X is chosen from 6 and 9 and Y and Z are chosen from 0, 1, and 8 is not detour-prone. There are $2 \cdot 3^2 = 18$ such ZIP codes. Any ZIP code of the form XYZY′X where Y is chosen from 6 and 9 and X and Z are chosen from 0, 1, and 8 is

not detour-prone. There are $2 \cdot 3^2 = 18$ such ZIP codes. Any ZIP code of the form XYZY′X′ where X and Y are chosen from 6 and 9 and Z are chosen from 0, 1, and 8 is not detour-prone. There are $2^2 \cdot 3 = 12$ such ZIP codes.
Number of detour-prone ZIP codes:
$3125 - 27 - 18 - 18 - 12 = 3050$ ZIP codes

37. Committees with no restrictions:
$C(12, 4) = 495$ committees
Committees with no seniors:
$C(9, 4) = 126$ committees
Committees with at least one senior:
$495 - 126 = 369$ committees

39. Consider the case with 3 tap, 1 ballet, and 1 modern routine. There are
$C(8, 3) \cdot C(5, 1) \cdot C(2, 1) = 56 \cdot 5 \cdot 2$
$= 560$ ways to do this.
Consider the case with 1 tap 3 ballet, and 1 modern routine. There are
$C(8, 1) \cdot C(5, 3) \cdot C(2, 1) = 8 \cdot 10 \cdot 2$
$= 160$ ways to do this.
Consider the case with 2 tap, 2 ballet, and 1 modern routine. There are
$C(8, 2) \cdot C(5, 2) \cdot C(2, 1) = 28 \cdot 10 \cdot 2$
$= 560$ ways to do this.
Consider the case with 2 tap, 1 ballet, and 2 modern routines. There are
$C(8, 2) \cdot C(5, 1) \cdot C(2, 2) = 28 \cdot 5 \cdot 1$
$= 140$ ways to do this.
Consider the case with 1 tap, 2 ballet, and 2 modern routines. There are
$C(8, 1) \cdot C(5, 2) \cdot C(2, 2) = 8 \cdot 10 \cdot 1$
$= 80$ ways to do this.
Thus the manager can choose
$560 + 160 + 560 + 140 + 80 = 1500$ ways.

41. First order the men. There are $6! = 720$ ways to do this. Then order the women. There are $6! = 720$ ways to do this. Since no two women can sit next to each other, there must be at least one man between each two. This is a total of five men. The sixth man can be at the end or with one of the other men. There are 7 ways to do this. Thus there are a total of $720 \cdot 720 \cdot 7 = 3,628,800$ ways to seat them.

43. Each family member can decide to come or not come to dinner. Thus there are $2^6 = 64$ different groups that can come to dinner.

45. 3 blue marbles and no white marbles chosen:
$C(3, 3) \cdot C(5, 0) = 1 \cdot 1 = 1$ way
2 blue marbles and 1 white marble chosen:
$C(3, 2) \cdot C(5, 1) = 3 \cdot 5 = 15$ ways
number of blue marbles chosen exceeds the number of white marbles chosen:
$1 + 15 = 16$ ways

47. Total number of possible bridge hands:
$C(52, 13)$
Bridge hands with all four aces: $C(48, 9)$
Percentage of bridge hands with all four aces: $\dfrac{C(48, 9)}{C(52, 13)} \approx .00264 = .264\%$

49. Bridge hands with all four aces:
$C(48, 9) = 1,677,106,640$
Bridge hands with the two red kings, the two red queens, and no other kings or queens:
$C(44, 9) = 708,930,508$
A bridge hand with all four aces is more likely.

Exercises 7

1. $\dbinom{6}{2} = C(6, 2) = \dfrac{6 \cdot 5}{2 \cdot 1} = 15$

3. $\dbinom{8}{1} = C(8, 1) = \dfrac{8}{1} = 8$

5. $\dbinom{18}{16} = C(18, 16) = C(18, 2) = \dfrac{18 \cdot 17}{2 \cdot 1} = 153$

7. $\dbinom{7}{0} = C(7, 0) = 1$

9. $\dbinom{8}{8} = C(8, 8) = 1$

11. $\dbinom{n}{n-1} = C(n, n-1) = C(n, 1) = \dfrac{n}{1} = n$

13. $0! = 1$

15. $n \cdot (n-1)! = n!$

17. $\dbinom{6}{0} + \dbinom{6}{1} + \dbinom{6}{2} + \dbinom{6}{3} + \dbinom{6}{4} + \dbinom{6}{5} + \dbinom{6}{6} = 2^6 = 64$

19. $\dbinom{10}{0}x^{10} + \dbinom{10}{1}x^9 y + \dbinom{10}{2}x^8 y^2 = x^{10} + 10x^9 y + 45x^8 y^2$

21. $\dbinom{15}{13}x^2 y^{13} + \dbinom{15}{14}xy^{14} + \dbinom{15}{15}y^{15} = 105x^2 y^{13} + 15xy^{14} + y^{15}$

23. $\dbinom{20}{10}x^{10}y^{10} = 184,756x^{10}y^{10}$

25. $\dbinom{11}{7} = 330$

27. $2^6 = 64$ subsets

29. $2^4 = 16$ tips

31. $2^6 = 64$ options

33. $2^8 - 1 = 255$ ways

35. $2 \cdot 3 \cdot 2^{15} = 196,608$ types

37. $2^7 - C(7, 6) - C(7, 7) = 128 - 7 - 1 = 120$ ways

39. Find the number of subsets of the set $\{a, b, d, e\}$.
$2^4 = 16$ subsets

41. $2^8 - C(8, 0) - C(8, 1) = 256 - 1 - 8 = 247$ ways

43. $2^{12} = 4096$ groups

45. Each region either contains elements in no sets, exactly one set, exactly two sets, or all three sets. The number of regions containing elements in no set is $\dbinom{3}{0}$. The number of regions containing elements in exactly one set is $\dbinom{3}{1}$. The number of regions containing elements in exactly two sets is $\dbinom{3}{2}$. The number of regions containing elements in all three sets is $\binom{3}{3}$. Thus there are

$$\binom{3}{0} + \binom{3}{1} + \binom{3}{2} + \binom{3}{3} = 1 + 3 + 3 + 1 = 8 \text{ regions.}$$

47. $C(8, 3) \cdot C(12, 2) = 56 \cdot 66 = 3696$ ways

49. $(x - 3y)^7 = [x + (-3y)]^7$

    The term with y^4 is $\binom{7}{4}x^3(-3y)^4 = 35x^3 \cdot 81y^4 = 2835x^3y^4$. The coefficient of y^4 is $2835x^3$.

51. $0 = 0^n = (1-1)^n$

    $$= \binom{n}{0}1^n + \binom{n}{1}1^{n-1} \cdot (-1) + \binom{n}{2}1^{n-2} \cdot (-1)^2 + \cdots + \binom{n}{n-1}1 \cdot (-1)^{n-1} + \binom{n}{n}(-1)^n$$

    $$= \binom{n}{0} - \binom{n}{1} + \binom{n}{2} + \cdots \pm \binom{n}{n}$$

53. From Exercise 52, $\binom{n}{k-1} + \binom{n}{k} = \binom{n+1}{k}$.

    Thus the kth entry in the $(n + 1)$st row is the sum of the $(k - 1)$st and kth entry in the nth row. The first and last entry of each row is 1.

1	1							
1	2	1						
1	3	3	1					
1	4	6	4	1				
1	5	10	10	5	1			
1	6	15	20	15	6	1		
1	7	21	35	35	21	7	1	
1	8	28	56	70	56	28	8	1

55. The feasible sets must include the inequalities $x \geq y - 3$ and $x + 3y \leq 21$. Thus the feasible set includes subsets of the set of the three remaining inequalities.

    $2^3 = 8$ feasible sets

Exercises 8

1. $\dfrac{5!}{3!1!1!} = 20$

3. $\dfrac{6!}{2!1!2!1!} = 180$

5. $\dfrac{7!}{3!2!2!} = 210$

7. $\dfrac{12!}{4!4!4!} = 34,650$

9. $\dfrac{12!}{5!3!2!2!} = 166,320$

11. $\dfrac{1}{5!} \cdot \dfrac{15!}{(3!)^5} = 1,401,400$

13. $\dfrac{1}{3!} \cdot \dfrac{18!}{(6!)^3} = 2,858,856$

15. $\begin{pmatrix} 20 \\ 7,\, 5,\, 8 \end{pmatrix} = \dfrac{20!}{7!5!8!} = 99,768,240$ reports

17. $\begin{pmatrix} 30 \\ 10,\, 2,\, 18 \end{pmatrix} = \dfrac{30!}{10!2!18!}$
     $= 5,708,552,850$ ways

19. $\dfrac{1}{4!} \cdot \dfrac{20!}{(5!)^4} = 488,864,376$ ways

21. $\begin{pmatrix} 4 \\ 1,\, 1,\, 2 \end{pmatrix} = \dfrac{4!}{1!1!2!} = 12$ ways

23. $\dfrac{1}{7!} \cdot \dfrac{14!}{(2!)^7} = 135,135$ ways

25. Select the number of elements of S_1:
     $$p_1 = \begin{pmatrix} n \\ n_1 \end{pmatrix} = \dfrac{n!}{n_1!(n-n_1)!} \text{ ways}$$
     Select the number of elements of S_2:

$$p_2 = \binom{n - n_1}{n_2} = \frac{(n - n_1)!}{n_2 \,!(n - n_1 - n_2)!} \text{ ways}$$

Select the number of elements of $S_k (1 < k < m)$:

$$p_k = \binom{n - n_1 - n_2 - \cdots - n_{k-1}}{n_k} = \frac{(n - n_1 - n_2 - \cdots - n_{k-1})!}{n_k \,!(n - n_1 - n_2 - \cdots - n_{k-1} - n_k)!} \text{ ways}$$

Select the number of elements of S_m:

$$p_m = \binom{n - n_1 - n_2 - \cdots - n_{m-1}}{n_m} = \binom{n_m}{n_m} = \frac{n_m\,!}{n_m\,!} \text{ ways}$$

Observe that for $1 < k \le m$, the numerator of p_k is $(n - n_1 - \cdots - n_{k-1})!$ which also occurs in the denominator of p_{k-1}, which is $n_{k-1}\,!(n - n_1 - \cdots - n_{k-1})!$.

Using the generalized multiplication principle, the number of ordered partitions is $p_1 \cdot p_2 \cdots p_m$. All of the numerators cancel in this multiplication for $1 < k \le m$, leaving $n_1\,! \cdot n_2\,! \cdots n_m\,!$ in the denominator. Thus, $f_1\,! \cdot f_2\,! \cdots p_m = \dfrac{n!}{n_1\,! \cdot n_2\,! \cdot n_3\,! \cdots n_m\,!}$.

27. $\dfrac{1}{5!} \cdot \dfrac{65!}{(13!)^5} \approx 8.81 \times 10^{41}$ ways

Chapter 5 Supplementary Exercises

1. $\varnothing, \{a\}, \{b\}, \{a, b\}$

2. $(S \cup T')' = S' \cap T$

3. $C(16, 2) = \dfrac{16!}{2!14!} = 120$ possibilities

4. $2 \cdot 5! = 240$ ways

5.

6. $\dbinom{12}{0} x^{12} + \dbinom{12}{1} x^{11} y + \dbinom{12}{2} x^{10} y^2 = x^{12} + 12 x^{11} y + 66 x^{10} y^2$

7. $C(8, 3) \cdot C(6, 2) = \dfrac{8!}{3!5!} \cdot \dfrac{6!}{2!4!} = 56 \cdot 15 = 840$

8. Let $U = \{\text{people}\}$, $P = \{\text{people who received placebos}\}$, $I = \{\text{people who showed improvement}\}$.
$n(U) = 60$; $n(P) = 15$; $n(I) = 40$; $n(P' \cap I) = 30$
Draw and complete a Venn diagram as shown.

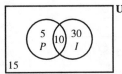

$n(P' \cap I') = 15$

Fifteen of the people who received the drug showed no improvement.

9. $7 \cdot 5 = 35$ combinations

10. $\begin{pmatrix} 12 \\ 2, 4, 6 \end{pmatrix} = \dfrac{12!}{2!4!6!} = 13,860$

11. Let $U = \{\text{applicants}\}$, $F = \{\text{applicants who speak French}\}$, $S = \{\text{applicants who speak Spanish}\}$, and $G = \{\text{applicants who speak German}\}$.
$n(U) = 115; n(F) = 70; n(S) = 65; n(G) = 65;$
$n(F \cap S) = 45; n(S \cap G) = 35; n(F \cap G) = 40; n(F \cap S \cap G) = 35$
Draw and complete a Venn diagram as shown.

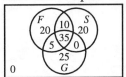

$n((F \cup S \cup G)') = 0$

None of the people speak none of the three languages.

12. $\begin{pmatrix} 17 \\ 15 \end{pmatrix} = \begin{pmatrix} 17 \\ 2 \end{pmatrix} = \dfrac{17 \cdot 16}{2 \cdot 1} = 136$

For Exercises 13–20, let $U = \{\text{members of the Earth Club}\}$,
$W = \{\text{members who thought the priority is clean water}\}$, $A = \{\text{members who thought the priority is clean air}\}$,
$R = \{\text{members who thought the priority is recycling}\}$. Then $n(U) = 100, n(W) = 45, n(A) = 30, n(R) = 42,$
$n(W \cap A) = 13, n(A \cap R) = 20, n(W \cap R) = 16, n(W \cap A \cap R) = 9$. Draw and complete the Venn diagram as follows.

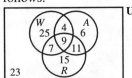

13. $n(A \cap W' \cap R') = 6$

14. $n((W \cap A') \cup (W' \cap A)) = (25 + 7) + (6 + 11) = 32 + 17 = 49$

15. $n((W \cup R) \cap A') = 25 + 15 + 7 = 47$

16. $n(A \cap R \cap W') = 11$

17. $n((W \cap A' \cap R') \cup (W' \cap A \cap R') \cup (W' \cap A' \cap R)) = 25 + 6 + 15 = 46$

18. $n(R') = 23 + 25 + 6 + 4 = 58$

19. $n(R \cap A') = 15 + 7 = 22$

20. $n((W \cup A \cup R)') = 23$

21. $C(9, 4) = C(9, 5) = 126$

22. 2^{40}

23. $S = \{$students who ski$\}$, $H = \{$students who play ice hockey$\}$
$n(S \cup H) = n(S) + n(H) - n(S \cap H) = 400 + 300 - 150 = 550$

24. $6 \cdot 10 \cdot 8 = 480$ meals

25. $\begin{pmatrix} 5 \\ 1, 3, 1 \end{pmatrix} = \dfrac{5!}{1!3!1!} = 20$ ways

26. The first digit can be anything but 0, the hundreds digit must be 3, and the last digit must be even.
$9 \cdot 1 \cdot 5 \cdot 10^4 = 450,000$

27. $9^2 \cdot 10^8 = 8,100,000,000$

28. $26 + 26^2 + 26^3 + \cdots + 26^{10} = 146,813,779,479,510$

29. Strings of length 8 formed from the symbols a, b, c, d, e:
$5^8 = 390,625$ strings
Strings of length 8 formed from the symbols a, b, c, d:
$4^8 = 65,536$ strings
Strings with at least one e:
$390,625 - 65,536 = 325,089$ strings

30. $C(12, 5) = \dfrac{12!}{5!7!} = 792$

31. $C(100, 14) = \dfrac{100!}{14!86!} = 44,186,942,677,323,600$ groups

32. $U = \{$households$\}$
$F = \{$households that get *Fancy Diet Magazine*$\}$
$C = \{$households that get *Clean Living Journal*$\}$
$n(F \cup C) = n(F) + n(C) - n(F \cap C) = 4000 + 10,000 - 1500 = 12,500$
$n((F \cup C)') = n(U) - n(F \cup C) = 40,000 - 12,500 = 27,500$
27,500 households get neither.

33. $3^{10} = 59,049$ paths

34. $C(60, 10) = \dfrac{60!}{10!50!} = 75,394,027,566$ ways

35. $5^{10} = 9,765,625$ tests

36. $21^6 = 85,766,121$ strings

37. $C(10, 4) = \dfrac{10!}{4!6!} = 210$ ways

38. $\dfrac{1}{3!} \cdot \dfrac{21!}{(7!)^3} = 66,512,160$ ways

39. $14! = 87,178,291,200$ ways

40. $\dfrac{1}{5!} \cdot \dfrac{100!}{(20!)^5}$ ways

41. Suppose one of the senators from New York has been assigned to a group. There are $C(98, 19)$ ways to complete this group.

$$C(98, 19) = \frac{98!}{19!79!} = \frac{20 \cdot 80 \cdot 98!}{20!80!}$$

To assign the remaining 80 senators into groups of 20 each, there are $\dfrac{1}{4!} \cdot \dfrac{80!}{(20!)^4}$ ways.

Thus the number of ways to divide the senators if the 2 senators from the state of New York cannot be in the same group is as follows.

$$C(98, 19) \cdot \frac{1}{4!} \cdot \frac{80!}{(20!)^4} = \left(\frac{20 \cdot 80 \cdot 98!}{20!80!} \right)\left(\frac{1}{4!} \frac{80!}{(20!)^4} \right)$$

$$= \frac{20 \cdot 80 \cdot 98!}{4!(20!)^5} = \frac{5 \cdot 20 \cdot 80 \cdot 98!}{5 \cdot 4!(20!)^5} = \frac{100 \cdot 80 \cdot 98!}{5!(20!)^5}$$

$$= \frac{80 \cdot 100 \cdot 99 \cdot 98!}{99 \cdot 5!(20!)^5} = \frac{80 \cdot 100!}{99 \cdot 5!(20!)^5} = \frac{80}{99} \cdot \frac{1}{5!} \cdot \frac{100!}{(20!)^5}$$

Thus there are $\dfrac{80}{99} \cdot \dfrac{1}{5!} \cdot \dfrac{100!}{(20!)^5}$ ways to divide the senators.

$\left(\text{This is } \dfrac{80}{99} \text{ of the answer in Exercise 40.} \right)$

42. To form a palindrome with only one letter, choose one letter. There are $C(10, 1) = 10$ ways to do this. To form a palindrome using two different letters, choose two letters and then decide which is the middle letter. There are $C(10, 2) \cdot 2 = 45 \cdot 2 = 90$ ways to do this.
Thus, there are $10 + 90 = 100$ palindromes.

43. To form a palindrome with only one letter, choose one letter. There are $C(10, 1) = 10$ ways to do this. To form a palindrome using two different letters, choose two letters and then decide which forms the middle two letters. There are $C(10, 2) \cdot 2 = 45 \cdot 2 = 90$ ways to do this.
Thus, there are $10 + 90 = 100$ palindromes.

44. To form a palindrome with only one letter, choose one letter. There are $C(10, 1) = 10$ ways to do this. To form a palindrome using two different letters, choose two letters and then decide on the arrangement for the first three letters of the palindrome. There are $2^3 - 2 = 6$ ways to arrange the first three letters of the palindrome, so there are $C(10, 2) \cdot 6 = 45 \cdot 6 = 270$ ways to use 2 letters. To form a palindrome using three different letters, choose three letters and then decide on the arrangement for the first three letters of the palindrome. There are $3! = 6$ ways to arrange the first three letters of the palindrome, so there are $C(10, 3) \cdot 6 = 120 \cdot 6 = 720$ ways to use 3 letters. Thus, there are $10 + 270 + 720 = 1000$ palindromes.

45. $5! \cdot 4! \cdot 3! \cdot 2! \cdot 1! = 120 \cdot 24 \cdot 6 \cdot 2 \cdot 1 = 34{,}560$

46. $P(12, 5) = 12 \cdot 11 \cdot 10 \cdot 9 \cdot 8 = 95{,}040$

47. There are four suits. Once a suit is chosen, select 5 of the 13 cards.
Poker hands with cards of the same suit:
$4 \cdot C(13, 5) = 4 \cdot 1287 = 5148$ hands

48. $9 \cdot 9 \cdot 8 = 648$ numbers

49. Exactly the first two digits alike: $9 \cdot 1 \cdot 9 = 81$
First and last digit alike: $9 \cdot 9 \cdot 1 = 81$
Last two digits alike: $9 \cdot 9 \cdot 1 = 81$
Numbers with exactly two digits alike: $81 + 81 + 81 = 243$

50. $C(4, 3) \cdot C(48, 2) = 4 \cdot 1128 = 4512$ hands

51. $24 \cdot 23 + 24 \cdot 23 \cdot 22 = 552 + 12{,}144 = 12{,}696$ names

52. $3! = 6$ pairings

53. $2 \cdot 3! \cdot 3! = 72$ ways

54. $3 \cdot 2 \cdot 2 \cdot 1 = 12$ ways

55. Any two lines will intersect, so the number of intersections is $C(10, 2) = 45$.
Since each of the ten lines is boundary there are ten sides to the feasible set and there are ten vertices. The number of intersections which occur outside the feasible set is $45 - 10 = 35$.

56. First teacher: $\dfrac{1}{4!} \cdot \dfrac{24!}{(6!)^4} = 96{,}197{,}645{,}544$ ways

Second teacher: $\dfrac{1}{6!} \cdot \dfrac{24!}{(4!)^6} = 4{,}509{,}264{,}634{,}875$ ways

The second teacher has more options.

57. $\begin{pmatrix} 10 \\ 3,\ 4,\ 3 \end{pmatrix} = \dfrac{10!}{3!4!3!} = 4200$

58. Let n be the number of books.
$n! = 120$, so $n = 5$. There are five books.

59. $7 \cdot 6 \cdot 5 \cdot 1 \cdot 4 \cdot 3 \cdot 2 \cdot 1 \cdot 1 = 5040$ orders

60. a. $C(12, 2)C(12, 3) = 66 \cdot 220 = 14{,}520$

b. Select five out of twelve couples. Then determine the spouse from each couple.
$C(12, 5) \cdot 2^5 = 25{,}344$

61. $C(7, 2) + C(7, 1) + C(7, 0) = 29$ ways

62. $2 \cdot 26 \cdot 26 \cdot 26 = 35{,}152$ call letters

63. Find n such that $n! = 479{,}001{,}600$.
Testing numbers on computer, $n = 12$.

64. $\begin{pmatrix} 25 \\ 10,\ 9,\ 6 \end{pmatrix} = 16{,}360{,}143{,}800$ ways

65. a. First choose the three letters. There are $C(26, 3)$ ways to do this. Then choose the three numbers. There are $C(10, 3)$ ways to do this. Then order the six. There are $6!$ ways to do this.
The number of license plates is $C(26, 3) \cdot C(10, 3) \cdot 6! = 224{,}640{,}000$.

b. The number of license plates consisting of three distinct letters followed by three distinct numbers is
$P(26, 3)P(10, 3) = C(26, 3) \cdot 3! \cdot C(10, 3) \cdot 3!$.
Thus there are $\dfrac{C(26, 3) \cdot C(10, 3) \cdot 6!}{C(26, 3) \cdot 3! \, C(10, 3) \cdot 3!} = 20$ times as many plates in (a).

Chapter 6

1. a. The set of all possible pairs: {RS, RT, RU, RV, ST, SU, SV, TU, TV, UV}

   b. The set of pairs containing R: {RS, RT, RU, RV}

   c. The set of pairs containing neither R nor S: {TU, TV, UV}

3. a. {HH, HT, TH, TT}

   b. {HH, HT}

5. a. {(I, red), (I, white), (II, red), (II, white)}

   b. All combinations with I: {(I, Red), (I, white)}

7. a. S = {all positive numbers of minutes}

   b. $E \cap F$ = "more than 5 but less than 8 minutes"
   $E \cap G = \varnothing$ (There's no time longer than 5 minutes but less than 4 minutes.)
   E' = "5 minutes or less"
   F' = "8 minutes or more"
   $E' \cap F = E'$ = "5 minutes or less"
   $E' \cap F \cap G = G$ = "less than 4 minutes"
   $E \cup F = S$

9. a. {(1, 1), (1, 2), (1, 3), (1, 4), (2, 1), (2, 2), (2, 3), (2, 4), (3, 1), (3, 2), (3, 3), (3, 4), (4, 1), (4, 2), (4, 3), (4, 4)}

   b. (i) {(2, 2), (2, 4), (4, 2), (4, 4)}
   (ii) {(1, 2), (1, 4), (2, 1), (2, 2), (2, 3), (2, 4), (3, 2), (3, 4), (4, 1), (4, 2), (4, 3), (4, 4)}
   (iii) {(3, 3), (3, 4), (4, 3), (4, 4)}
   (iv) {(2, 4), (3, 3), (4, 2)}
   (v) {(1, 4), (2, 3), (2, 4), (3, 2), (3, 3), (3, 4), (4, 1), (4, 2), (4, 3), (4, 4)}
   (vi) {(1, 1), (2, 2), (3, 3), (4, 4)}
   (vii) {(1, 2), (1, 3), (2, 1), (2, 2), (2, 4), (3, 1), (3, 3), (3, 4), (4, 2), (4, 3)}
   (viii) {(1, 1), (1, 2), (1, 3), (2, 1), (2, 2), (2, 3), (3, 1), (3, 2), (3, 3)}

11. a. No; $E \cup F = \{2\}$

   b. Yes; $F \cup G = \varnothing$

13. All combinations of members of S: $\varnothing$, $\{a\}$, $\{b\}$, $\{c\}$, $\{a, b\}$, $\{a, c\}$, $\{b, c\}$, S

15. Yes; $(E \cup F) \cap (E' \cap F') = \{1, 2, 3\} \cap \{4\} = \varnothing$

17. a. {0, 1, 2, 3, 4, 5, 6, 7, 8, 9, 10}

   b. More than half heads: {6, 7, 8, 9, 10}

19. **a.**   No; there are blue-eyed males.

    b.   Yes; a brown-eyed female doesn't have blue eyes.

    c.   Yes; a brown-eyed female isn't male.

21. Length of line in number of persons: the set of nonnegative integers.

23. Time in minutes: the set of nonnegative numbers.

25. The set of all possible triplets of three-digit numbers:
    {(000, 000, 000), (000, 000, 001), ... (999, 999, 999)}
    E' = the event that at least one number is odd
    $E \cap F$ = the event that all numbers are even and are more than 699

27. **a.**   6 suspects $\times$ 6 weapons $\times$ 9 rooms = 324 points

    b. $E \cap F$ = "The murder occurred in the library with a gun."

    c. $E \cup F$ = "Either the murder occurred in the library, or it was perpetrated with a gun."

Exercises 3

1. **a.** $\dfrac{46,277}{774,746}$

   b. $\dfrac{46,277 + 1855}{774,746} = \dfrac{48,132}{774,746}$

   c. $\dfrac{774,746 - 48,132}{774,746} = \dfrac{726,614}{774,746}$

3. **a.** E = "the numbers add up to 8" = {(2, 6), (3, 5), (4, 4), (5, 3), (6, 2)}
   $\Pr(E) = \dfrac{5}{36}$

   b. $\Pr(\text{sum is } 2) = \Pr((1, 1)) = \dfrac{1}{36}$; $\Pr(\text{sum is } 3) = \Pr((1, 2)) + \Pr((2, 1)) = \dfrac{2}{36}$

   $\Pr(\text{sum is } 4) = \Pr((1, 3)) + \Pr((2, 2)) + \Pr((3, 1)) = \dfrac{3}{36}$

   The probability that the sum is 2, 3, or 4 is $\dfrac{1}{36} + \dfrac{2}{36} + \dfrac{3}{36} = \dfrac{1}{6}$.

5. $\dfrac{1}{38} + \dfrac{1}{38} = \dfrac{2}{38} = \dfrac{1}{19}$

7. a. $1 - \left(\dfrac{1}{3} + \dfrac{1}{2} \right) = \dfrac{1}{6}$

  b. The probability is $\dfrac{1}{6}$. Thus $a = 1$ and $a + b = 6$, so $b = 5$. The odds are 1 to 5.

9. a. $.1 + .6 = .7$

  b. $.6 + .1 = .7$

11. a. $\dfrac{10}{10+1} = \dfrac{10}{11}$

  b. $\dfrac{1}{1+2} = \dfrac{1}{3}$

  c. $\dfrac{4}{4+5} = \dfrac{4}{9}$

13. $.09 = \dfrac{9}{100}$
Thus $a = 9$ and $a + b = 100$, so $b = 91$. The odds are 9 to 91.

15. Win: $\dfrac{11}{11+7} = \dfrac{11}{18}$
Lose: $\dfrac{7}{11+7} = \dfrac{7}{18}$

17. a. $\Pr(E \cup F) = \Pr(E) + \Pr(E) - \Pr(E \cap F) = .7$

  b. $\Pr(E \cap F') = \Pr(E) - \Pr(E \cap F) = .2$

19. a. $.20 + .25 + .25 = .7$

  b. $.70 \times 10{,}000 = 7000$

21.

Failures in:	Probability
Month 1	.05
Month 2	$.10 - \Pr(\text{first mo.}) = .10 - .05 = .05$
Month 3	$.20 - \Pr(\text{first 2 mos.}) = .20 - .10 = .10$
Month 4	$.25 - \Pr(\text{first 3 mos.}) = .25 - .20 = .05$
Month 5	$.30 - \Pr(\text{first 4 mos.}) = .30 - .25 = .05$
Month 6	$.32 - \Pr(\text{first 5 mos.}) = .32 - .30 = .02$
No failure in months 1–6	$1 - \Pr(\text{first 6 mos.}) = 1 - .32 = .68$

23. a. Pr(20–34) = .15
Pr(35–49) = Pr(20–49) – Pr(20–34) = .55
Pr(50–64) = Pr(20–64) – Pr(20–49) = .20
Pr(65–79) = Pr(20–79) – Pr(20–64) = .10

b. Pr(50–79) = Pr(20–79) – Pr(20–49) = .30

25. a. 100% – 17% = 83%

b. Some categories are left out—people who use a computer for both school and work, for example.

Exercises 4

1. a. There are $3 \times 3 \times 3 = 27$ combinations of class selections.
$$\frac{3}{27} = \frac{1}{9}$$

b. $\dfrac{3 \times 2 \times 1}{27} = \dfrac{2}{9}$

3. a. $\dfrac{7}{13}$

b. $\dfrac{6}{13}$

c. $\text{Pr}(\{3, 6, 9, 12\}) = \dfrac{4}{13}$

d. $\text{Pr}(\{1, 3, 5, 6, 7, 9, 11, 12, 13\} = \dfrac{9}{13}$

5. a. $\dfrac{C(6, 5)}{C(13, 5)} = \dfrac{6}{1287} = \dfrac{2}{429}$

b. $\dfrac{C(7, 5)}{C(13, 5)} = \dfrac{21}{1287} = \dfrac{7}{429}$

c. $1 - \dfrac{2}{429} = \dfrac{427}{429}$

7. $\dfrac{C(6, 2) \times C(5, 2)}{C(11, 4)} = \dfrac{15 \times 10}{330} = \dfrac{5}{11}$

9. Ways for exactly 2 to agree: $C(5, 2) \times C(4, 1)$
Ways for all 3 to agree: $C(5, 3)$
$$\frac{C(5, 2) \times C(4, 1) + C(5, 3)}{C(9, 3)} = \frac{10 \times 4 + 10}{84} = \frac{25}{42}$$

11. Ways for no girls to be chosen: $C(12, 7)$
Ways for exactly 1 girl to be chosen: $C(12, 6) \times C(10, 1)$
$$1 - \frac{C(12, 7) + C(12, 6) \times C(10, 1)}{C(22, 7)} = 1 - \frac{792 + 924 \times 10}{170,544} = \frac{16}{17}$$

13. Ways for no girl to be chosen: $C(12, 7)$
Ways for exactly 1 girl to be chosen: $C(12, 6) \times C(10, 1)$
Ways for exactly 2 girls to be chosen: $C(12, 5) \times C(10, 2)$
Ways for exactly 3 girls to be chosen: $C(12, 4) \times C(10, 3)$
$$\frac{C(12, 7) + C(12, 6) \times C(10, 1) + C(12, 5) \times C(10, 2) + C(12, 4) \times C(10, 3)}{C(22, 7)}$$
$$= \frac{792 + 9240 + 35,640 + 59,400}{170,544} = \frac{199}{323}$$

15. $1 - \dfrac{30 \times 29 \times 28 \times 27}{30^4} = \dfrac{47}{250}$

17. The probability of scoring 3, 4, or 5 equals the probability of scoring 0, 1, or 2: $\dfrac{1}{2}$.

19. **a.** $.2 + .15 - .1 = .25$

    b. $1 - .25 = .75$

    c. $1 - .2 = .8$

21. $\Pr(F) = 1 - \Pr(F') = .4$
$\Pr(E \cap F) = \Pr(E) + \Pr(F) - \Pr(E \cup F) = .3 + .4 - .7 = 0$

23. $\Pr(\text{second sock matches first}) = \dfrac{12 \cdot 1}{12 \cdot 11} = \dfrac{1}{11}$

25. There are $5! = 120$ ways to arrange the family.
First determine where the parents will stand. There are two ways with the man at one end. If the man doesn't stand at one end, there are $3 \cdot 2 = 6$ ways for the couple to stand together. Next determine the order of the children. There are $3! = 6$ ways to order the children.
Thus there are $(2 + 6) \cdot 6 = 48$ ways to stand with the parents together.
The probability is $\dfrac{48}{120} = \dfrac{2}{5}$.

27. The tourist must travel 8 blocks of which 3 are south. Thus he has $C(8, 3) = 56$ ways to get to B from A.

    a. To get from A to B through C there are $C(3, 1) \cdot C(5, 2) = 30$ ways.
    The probability is $\dfrac{30}{56} = \dfrac{15}{28}$.

b. To get from A to B through D there are $C(5, 1) \cdot C(3, 2) = 15$ ways.
The probability is $\dfrac{15}{56}$.

c. To get from A to B through C and D there are $C(3, 1) \cdot C(2, 0) \cdot C(3, 2) = 9$ ways.
The probability is $\dfrac{9}{56}$.

d. The number of ways to get from A to B through C or D is $30 + 15 - 9 = 36$.
The probability is $\dfrac{36}{56} = \dfrac{9}{14}$.

29. $1 - \dfrac{C(6, 3)}{C(10, 3)} = 1 - \dfrac{20}{120} = \dfrac{5}{6}$

31. $\dfrac{C(8, 2)}{C(9, 3)} = \dfrac{28}{84} = \dfrac{3}{9} = \dfrac{1}{3}$

33. $\dfrac{2}{C(40, 6)} = \dfrac{1}{1,919,190}$

35. a. $\dfrac{C(3, 2)}{C(11, 2)} = \dfrac{3}{55}$

b. 2 edges: 3 (A, C, F)
3 edges: 6 (D, E, G, H, I, J)
$\dfrac{3 \times 6}{C(11, 2)} = \dfrac{18}{55}$

c. Number of edge-sharing combinations
($AB, AH, BI, BJ, BC, CD, DJ, DE, EK, EF, FG, GK, GH, HI, IK, KJ$) ÷ Number of combinations
$= \dfrac{16}{C(11, 2)} = \dfrac{16}{55}$

d. $\dfrac{(\text{number on left}) \times (\text{number on right})}{C(11, 2)} = \dfrac{6}{11}$

37. a. Ways to guess exactly 8 right: $C(10, 8)$
Ways to guess exactly 9 right: $C(10, 9)$
Ways to guess all 10 right: $C(10, 10)$
$\dfrac{C(10, 8) + C(10, 9) + C(10, 10)}{2^{10}} = \dfrac{45 + 10 + 1}{1024} = \dfrac{7}{128}$

b.   Pr(not being labeled clairvoyant) $= \dfrac{121}{128}$

Pr(at least one in 10 being labeled clairvoyant) $= 1 - \left(\dfrac{121}{128}\right)^{10} \approx .43$

c. $\dfrac{121}{128}$

39. $1 - \dfrac{C(12,\ 3)}{C(15,\ 3)} = 1 - \dfrac{220}{455} = \dfrac{47}{91} \approx .5165$

41. $1 - \dfrac{3 \cdot 3 \cdot 2}{4 \cdot 4 \cdot 3} = 1 - \dfrac{3}{8} = \dfrac{5}{8} = .625$

43.   Answers will vary.

45.   **a.** $\dfrac{1}{6} \times 24 = 4$

      b. $\dfrac{C(24,\ 4) \times 5^{20}}{6^{24}} \approx .2139$

47. $1 - \dfrac{N \times (N-1) \times (N-2) \times \cdots \times (N - 20 + 1)}{N^{20}}$ dips below .5 starting with $N = 281$.

Exercises 5

1. $\Pr(E|F) = \dfrac{.1}{.3} = \dfrac{1}{3}, \ \Pr(F|E) = \dfrac{.1}{.5} = \dfrac{1}{5}$

3. $\Pr(E|F') = \dfrac{\Pr(E \cap F')}{\Pr(F')} = \dfrac{P(E) - \Pr(E \cap F)}{1 - \Pr(F)} = \dfrac{.4}{.7} = \dfrac{4}{7}$

5.   No; $\Pr(\text{has cancer}|\text{works for Ajax}) \neq \Pr(\text{has cancer})$

7.   **a.** $.80 \times .75 \times .60 = .36$

      b. $.36 + (1 - .80)(.75)(.60) + (.80)(1 - .75)(.60) + (.80)(.75)(1 - .60) = .81$

9. $(1 - .01)^5 (1 - .02)^5 (1 - .025)^3 = (.99)^5 (.98)^5 (.975)^3 \approx .7967$

11. $\Pr(E|F) = \dfrac{\frac{1}{4}}{\frac{1}{3}} = \dfrac{3}{4}$

      $\Pr(F|E) = \dfrac{\frac{1}{4}}{\frac{1}{2}} = \dfrac{1}{2}$

13. a. $1 - \text{Pr(Dem. or favors)} = 1 - [\text{Pr(Dem.)} + \text{Pr(favors)} - \text{Pr(Dem. and favors)}] = .4$

  b. $\dfrac{\text{Pr(Dem. and favors)}}{\text{Pr(Dem.)}} = .6$

  c. $\dfrac{\text{Pr(Dem. and favors)}}{\text{Pr(favors)}} = .75$

15. $\dfrac{\text{Pr(\{HHH\})}}{\text{Pr(\{HHH, HHT, HTH, THH\})}} = \dfrac{1}{4}$

17. $1 - \text{Pr(Neither } A \text{ nor } B \text{ lives 15 more years)} = 1 - (1 - .8)(1 - .7) = .94$

19. $(1 - .6)(1 - .6) = .16$

21. $1 - (1 - .005)^2 = .009975$

23. $\dfrac{15^4}{15 + 15^2 + 15^3 + 15^4} = \dfrac{5065}{54,240} \approx .9334$

25. $\dfrac{1}{15}$

27.   First consider the case where the codeword starts with a letter.
$5 \cdot 10 + 5 \cdot 10 \cdot 5 + 5 \cdot 10 \cdot 5 \cdot 10 = 2800$
Next consider the case where the codeword starts with a number.
$10 \cdot 5 + 10 \cdot 5 \cdot 10 + 10 \cdot 5 \cdot 10 \cdot 5 = 3050$
Thus there are $2800 + 3050 = 5850$ codewords.

29. $(.7)^4 = .2401$

31. $\Pr(E \cap F) = \Pr(E) \cdot \Pr(F)$ Definition of "independent"
$\Pr(E) + \Pr(F) - \Pr(E \cup F) = \Pr(E) \cdot \Pr(F)$ Substitution
$[1 - \Pr(E')] + [1 - \Pr(F')] - [1 - \Pr(E' \cap F')] = [1 - \Pr(E')] \cdot [1 - \Pr(F')]$ Substitution
$\Pr(E' \cap F') = \Pr(E') \cdot \Pr(F')$ Simplification

33.   0 points: $1 - .6 = .4$
1 point: $.6 \times .4 = .24$
2 points: $.6 \times .6 = .36$

35. $\Pr(E \cup F | G) = \dfrac{\Pr[(E \cup F) \cap G]}{\Pr(G)} = \dfrac{\Pr[(E \cap G) \cup (F \cap G)]}{\Pr(G)}$
$= \dfrac{\Pr(E \cap G) + \Pr(F \cap G) - \Pr[(E \cap G) \cap (F \cap G)]}{\Pr(G)}$

$$= \frac{\Pr(E \cap G)}{\Pr(G)} + \frac{\Pr(F \cap G)}{\Pr(G)} - \frac{\Pr[(E \cap F) \cap G]}{\Pr(G)}$$
$$= \Pr(E|G) + \Pr(F|G) - \Pr(E \cap F|G)$$

37. a. $\Pr\big(\text{death}|A\big) = \dfrac{12,000}{120,000} = \dfrac{1}{10}$

     b. $1000 \times \dfrac{1}{10} = 100$ per 1000

     c. $\Pr\big(\text{death}|B\big) = \dfrac{4500}{90,000} = \dfrac{1}{20}$

39. a. $\Pr(B) = 80\% \times (100\% - 65\%) = .28$

     b. $\Pr(C) = 20\% \times (100\% - 65\%) = .07$

     c. $\Pr(H) = (65\% \times 75\%) + (28\% \times 40\%) + (7\% \times 10\%) = .6065$

41. a. $\dfrac{.326 \times .113}{(.021 \times .230) + (.114 \times .250) + (.113 \times .196) + (.326 \times .113)} \approx .399$

     b. $(.021 \times .230) + (.114 \times .250) + (.113 \times .196) + (.3267 \times .113) = .092316 \approx 9.2\%$

43.   Total students $= 130 + 460 + 210 + 100 + \cdots + 50 = 2700$

     a. $\dfrac{200 + 300 + 50}{2700} \approx .2037$

     b. $\dfrac{130 + 100 + 80 + 200}{2700} \approx .1889$

     c. $\dfrac{460}{130 + 460 + 210} = .5750$

     d. $\dfrac{210}{210 + 150 + 100 + 50} \approx .4118$

     e. $\dfrac{100 + 50}{210 + 150 + 100 + 50} \approx .2941$

     f. $\dfrac{100 + 500 + 80 + 420}{100 + 500 + 150 + 80 + 420 + 100} \approx .8148$

45.   Total population $= 1000 + 1200 + \cdots + 10,000 = 147,250$

   a. $\dfrac{100,000+35,000+10,000}{147,250} \approx .985$

   b. $\dfrac{1000+1200+50}{147,250} \approx .015$

   c. $\dfrac{1200}{1200+35,000} \approx .033$

   d. $\dfrac{1000+50}{1000+100,000+50+10,000} \approx .009$

   e. No; Pr(has virus|Afro-American) $\neq$ Pr(has virus)

47. Less than 70 and divisible by 7: 7, 14, 21, 28, 35, 42, 49, 56, 63. Less than 70 and divisible by 3: 3, 6, 9, 12, 15, 18, 21, 24, 27, 30, 33, 36, 39, 42, 45, 48, 51, 54, 57, 60, 63, 66, 69. In both groups: 21, 42, 63.

Pr(both choose 21) = Pr(both choose 42) = Pr(both choose 63) $= \dfrac{1}{9} \times \dfrac{1}{23} = \dfrac{1}{207}$.

Pr(choose same) $= \dfrac{1}{207} + \dfrac{1}{207} + \dfrac{1}{207} = \dfrac{1}{69}$

49. Pr(all red) $= \dfrac{C(6,\,3)}{C(13,\,3)} = \dfrac{10}{143}$

Pr(all blue) $= \dfrac{C(4,\,3)}{C(13,\,3)} = \dfrac{2}{143}$;   Pr(all white) $= \dfrac{C(3,\,3)}{C(13,\,3)} = \dfrac{1}{286}$

Pr(all same color) = Pr(all red) + Pr(all blue) + Pr(all white) $= \dfrac{10}{143} + \dfrac{2}{143} + \dfrac{1}{286} = \dfrac{25}{286} \approx .0874$

51. Pr(both roast beef) = Pr(both ham) $= \dfrac{C(2,\,2)}{C(4,\,2)} = \dfrac{1}{6}$

Pr(both roast beef) + Pr(both ham) $= \dfrac{1}{6} + \dfrac{1}{6} = \dfrac{1}{3}$

53. $.650(1 - .650) = .2275$

55. $\Pr(W) = \dfrac{53,010}{117,378} \approx .452$

$\Pr(U) = \dfrac{5725}{117,378} \approx .049$

$\Pr(W \cap U) = \dfrac{2555}{117,378} \approx .022$

Not independent; $\Pr(W \cap U) \neq \Pr(W) \times \Pr(U)$

57. a. $\dfrac{4}{52} \cdot \dfrac{3}{51} \cdot \dfrac{2}{50} \cdot \dfrac{1}{49} \approx 3.69 \times 10^{-6}$

b. $\dfrac{2}{52} \cdot \dfrac{1}{51} \cdot \dfrac{2}{50} \cdot \dfrac{1}{49} \approx 6.16 \times 10^{-7}$

c. (a)

Exercises 6

1.

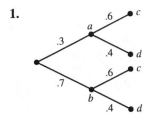

3.

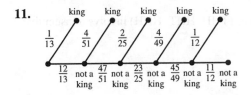

5. $(1 - .8)(1 - .6) = .08$

7. $.10 + .30 \times .05 + .60 \times .30 = .295$

9. $\text{Pr(white, then red)} + \text{Pr(red, then red)} = \dfrac{2}{3} \times \dfrac{1}{2} + \dfrac{1}{3} \times \dfrac{3}{4} = \dfrac{7}{12}$

11.

(tree diagram with branches labeled king / not a king)

$\dfrac{1}{13}$ $\dfrac{4}{51}$ $\dfrac{2}{25}$ $\dfrac{4}{49}$ $\dfrac{1}{12}$

$\dfrac{12}{13}$ $\dfrac{47}{51}$ $\dfrac{23}{25}$ $\dfrac{45}{49}$ $\dfrac{11}{12}$

Pr(king on 1st draw) + Pr(king on 2nd draw) + Pr(king on 3rd draw) = 1 − Pr(not a king on 3rd draw)

$= 1 - \dfrac{12}{13} \times \dfrac{47}{51} \times \dfrac{23}{25} = 1 - \dfrac{4324}{5525} = \dfrac{1201}{5525} \approx .22$

13. $\text{Pr(math final)} = \dfrac{1}{3} \times .20 + \dfrac{1}{3} \times .10 + \dfrac{1}{3} \times .05 = \dfrac{7}{60} \approx .1167$

$\text{Pr(biology final)} = \dfrac{1}{2} \times .20 + \dfrac{1}{2} \times .13 = \dfrac{33}{200} = .165$

$\text{Pr(math final)} + \text{Pr(biology final)} - \text{Pr(final in both)} = \dfrac{7}{60} + \dfrac{33}{200} - \dfrac{7}{60} \times \dfrac{33}{200} \approx .262$

15. $\Pr(\text{male}|\text{color-blind}) = \dfrac{\frac{1}{2} \times .05}{\frac{1}{2} \times .05 + \frac{1}{2} \times .004} = \dfrac{25}{27}$

17. $(.5)(.9)(.9) + (.5)(.1)(.7) + (.5)(.7)(.9) + (.5)(.3)(.7) = .86$

19. $\Pr\left(\text{fake}|HH\right) = \dfrac{\frac{1}{4} \times 1}{\frac{3}{4} \times \frac{1}{4} + \frac{1}{4} \times 1} = \dfrac{4}{7}$

21. $\Pr\left(\text{grove I}|\text{grapefruit}\right) = \dfrac{1000}{1000 + 1000} = \dfrac{1}{2}$

23. $1 - (.9999)^n$

25. Same shape
    $\Pr(\text{winning}) = \dfrac{3}{6} + \dfrac{3}{6} \times \dfrac{3}{5} \times \dfrac{2}{4} + \dfrac{3}{6} \times \dfrac{3}{5} \times \dfrac{2}{4} \times \dfrac{2}{3} \times \dfrac{1}{2} + \dfrac{3}{6} \times \dfrac{2}{5} \times \dfrac{3}{4} \times \dfrac{2}{3} \times \dfrac{1}{2} = \dfrac{3}{4}$
    Probability of winning card game is greater.

27. **a.** $\Pr\left(\text{after Cleveland}|\text{childless}\right) = \dfrac{1}{1 + 5} = \dfrac{1}{6}$

    b. $\Pr\left(\text{after Cleveland}|\text{son}\right) = \dfrac{32}{32 + 56} = \dfrac{32}{88} = \dfrac{4}{11}$

29. $\Pr(\text{night}|\text{part-timer}) = \dfrac{.60 \times 2}{.25 \times 4 + .60 \times 2} = \dfrac{6}{11}$

31. For three tosses, there are $2 \times 2 \times 2 = 8$ combinations. Three (THH, HHT, HHH) involve consecutive heads.
    $\Pr(\text{no consecutive heads}) = \dfrac{8 - 3}{8} = \dfrac{5}{8}$
    For four tosses, there are $2^4 = 16$ combinations.
    Eight (TTHH, THHT, HHTT, THHH, HTHH, HHTH, HHHT, HHHH) involve consecutive heads.
    $\Pr(\text{no consecutive heads}) = \dfrac{16 - 8}{16} = \dfrac{1}{2}$

Exercises 7

1. $\Pr\left(\text{over 60}|\text{accident}\right) = \dfrac{.10 \times .04}{(.05 \times .06) + (.10 \times .04) + (.25 \times .02) + (.20 \times .015) + (.30 \times .025) + (.10 \times .04)}$
   $= \dfrac{.004}{.0265} = \dfrac{8}{53}$

3. $\Pr\left(\text{sophomore}|A\right) = \dfrac{.30 \times .4}{(.10 \times .2) + (.30 \times .4) + (.40 \times .3) + (.20 \times .1)} = \dfrac{.12}{.28} = \dfrac{3}{7}$

5. $\Pr\left(\geq \$25,000 | 2t \text{ cars}\right) = \dfrac{.05 \times .9}{(.10 \times .2) + (.20 \times .5) + (.35 \times .6) + (.30 \times .75) + (.05 \times .9)} = \dfrac{.045}{.6}$

$= \dfrac{3}{40} = .075$

7. $\Pr\left(\text{guilty} | \text{left - handed}\right) = \dfrac{\Pr(\text{guilty}) \times \Pr\left(\text{left - handed} | \text{guilty}\right)}{\Pr(\text{guilty}) \times \Pr\left(\text{left - handed} | \text{guilty}\right) + \Pr(\text{innocent}) \times \Pr\left(\text{left - handed} | \text{innocent}\right)}$

$= \dfrac{.70 \times 1}{(.70 \times 1) + (.30 \times .20)} = \dfrac{35}{38} \approx 92\%$

9. a. $(.20 \times .20) + (.15 \times .15) + (.25 \times .12) + (.30 \times .10) + (.10 \times .10) = .1325$

    b. $\Pr\left(\text{division C} | \text{bilingual}\right) = \dfrac{.25 \times .12}{.1325} = \dfrac{.03}{.1325} = \dfrac{12}{53} \approx .23$

11. Let A = Al wins, B = Bob wins, C = Al randomized, D = Bob randomized, E = Al chose 20, and F = Bob chose 20.

$\Pr\left(A | B\right) = \dfrac{\Pr(A \cap B)}{\Pr(B)}$

$= \dfrac{\Pr(C \cap D) \times \Pr\left(A \cap B | C \cap D\right) + \Pr(E \cap F) \times \Pr\left(A \cap B | E \cap F\right) + \Pr(C \cap F) \times \Pr\left(A \cap B | C \cap F\right) + \Pr(D \cap E) \times \Pr\left(A \cap B | D \cap E\right)}{\Pr(B)}$

$= \dfrac{(.8 \times .8) \times \left(\frac{1}{99} \times \frac{1}{100}\right) + (.2 \times .2) \times \frac{1}{100} + (.8 \times .2) \times 0 + (.8 \times .2) \times 0}{\frac{1}{100}} = \dfrac{23}{495} \approx .046$

13. a. $\Pr(\text{one is}) = \dfrac{13}{52} = \dfrac{1}{4}$

    b. $\Pr\left(\text{none is} | \text{random one isn' t}\right)$

$= \dfrac{\Pr(\text{none is}) \times \Pr\left(\text{random one isn' t} | \text{none is}\right)}{\Pr(\text{none is}) \times \Pr\left(\text{random one isn' t} | \text{none is}\right) + \Pr(\text{one is}) \times \Pr\left(\text{random one isn' t} | \text{one is}\right)}$

$= \dfrac{\frac{3}{4} \times 1}{\left(\frac{3}{4} \times 1\right) + \left(\frac{1}{4} \times \frac{12}{13}\right)} = \dfrac{13}{17} \approx .765$

    c. $\Pr(\text{one is} | 10 \text{ randoms aren't})$

$= \dfrac{\Pr(\text{one is}) \times \Pr\left(10 \text{ randoms aren' t} | \text{one is}\right)}{\Pr(\text{one is}) \times \Pr\left(10 \text{ randoms aren' t} | \text{one is}\right) + \Pr(\text{none is}) \times \Pr\left(10 \text{ randoms aren' t} | \text{none is}\right)}$

$= \dfrac{\frac{1}{4} \times \left(\frac{12}{13}\right)^{10}}{\left[\frac{1}{4} \times \left(\frac{12}{13}\right)^{10}\right] + \left(\frac{3}{4} \times 1\right)} \approx .130$

15. a. $\text{Pr}\left(\text{Lakeside}|\text{winner}\right)$

$$= \frac{\text{Pr(Lakeside)} \times \text{Pr}\left(\text{winner}|\text{Lakeside}\right)}{\text{Pr(Lakeside)} \times \text{Pr}\left(\text{winner}|\text{Lakeside}\right) + \text{Pr(Pylesville)} \times \text{Pr}\left(\text{winner}|\text{Pylesville}\right) + \text{Pr(Millerville)} \times \text{Pr}\left(\text{winner}|\text{Millerville}\right)}$$

$$= \frac{.40 \times .05}{(.40 \times .05) + (.20 \times .02) + (.40 \times .03)} = \frac{5}{9}$$

b. $\dfrac{.20 \times .02}{(.40 \times .05) \times (.20 \times .02) + (.40 \times .03)} = \dfrac{1}{9} \approx 11\%$

17. a. $1 - .99 = .01$

b. $\text{Pr}\left(\text{pregnant}|\text{positive}\right)$

$$= \frac{\text{Pr(pregnant)} \times \text{Pr}\left(\text{positive}|\text{pregnant}\right)}{\text{Pr(pregnant)} \times \text{Pr}\left(\text{positive}|\text{pregnant}\right) + \text{Pr(not pregnant)} \times \text{Pr}\left(\text{positive}|\text{not pregnant}\right)}$$

$$= \frac{.40 \times .99}{(.40 \times .99) + (.60 \times .02)} = \frac{33}{34} \approx .971$$

19. $\text{Pr}\left(\text{steroids}|\text{positive}\right) = \dfrac{\text{Pr(steroids)} \times \text{Pr}\left(\text{positive}|\text{steroids}\right)}{\text{Pr(steroids)} \times \text{Pr}\left(\text{positive}|\text{steroids}\right) + \text{Pr(no steroids)} \times \text{Pr}\left(\text{positive}|\text{no steroids}\right)}$

$$= \frac{.10 \times .93}{(.10 \times .93) + (.90 \times .02)} = \frac{31}{37} \approx .838$$

21. $\text{Pr}\left(\text{cancer}|\text{positive}\right) = \dfrac{\text{Pr(cancer)} \times \text{Pr}\left(\text{positive}|\text{cancer}\right)}{\text{Pr(cancer)} \times \text{Pr}\left(\text{positive}|\text{cancer}\right) + \text{Pr(no cancer)} \times \text{Pr}\left(\text{positive}|\text{no cancer}\right)}$

$$= \frac{.02 \times .75}{(.02 \times .75) + (.98 \times .30)} = \frac{5}{103} \approx .049$$

Exercises 8

1. Use seq(int(6*rand)+1,X,1,72,1)$\rightarrow$L$_1$.

Theoretical probabilities: $\dfrac{1}{6}$ for each fall.

3. Use seq(int(100*rand)+1,X,1,10,1)$\rightarrow$L$_1$ where a number from 1–83 represents a successful freethrow and 84–100 represents a miss or seq(1–int(rand/.83),X,1,10,1)$\rightarrow$L$_1$ where 0 represents a successful freethrow and 1 represents a miss.

5. Use seq(int(4*rand)+1,X,1,10,1)$\rightarrow$L$_1$, where a = 1, b = 2, c = 3, and d = 4.

7. Answers will vary (see Example 4).

9. Answers will vary. Start by generating 13 random whole numbers from 1 to 10, where 1 unit of water = 1–4, 2 units = 5–7, 3 units = 8–9, and 4 units = 10.

Chapter 6 Supplementary Exercises

1. $1 - \left(\dfrac{1}{2}\right)^5 = \dfrac{31}{32}$

2. No; $\Pr(F|E) = \dfrac{3}{4} \neq \Pr(F) = \dfrac{1}{2}$

3. $\left(\dfrac{1}{3} \times \dfrac{2}{3}\right) + \left(\dfrac{2}{3} \times \dfrac{1}{3}\right) = \dfrac{4}{9}$

4. $\dfrac{5}{10} \times \dfrac{4}{9} \times \dfrac{3}{8} = \dfrac{1}{12}$

5. $\dfrac{\text{number of public colleges offering engineering majors}}{\text{number of public colleges}} = \dfrac{15 - 5}{50 - 25} = \dfrac{2}{5}$

6. $\Pr(\text{correct}|\text{rejected}) = \dfrac{\Pr(\text{correct}) \times \Pr(\text{rejected}|\text{correct})}{\Pr(\text{correct}) \times \Pr(\text{rejected}|\text{correct}) + \Pr(\text{incorrect}) \times \Pr(\text{rejected}|\text{incorrect})}$

   $= \dfrac{.80 \times .05}{(.80 \times .05) + (.20 \times .90)} = \dfrac{2}{11}$

7. $\dfrac{20}{100} = \dfrac{1}{5}$

8. $\dfrac{2000}{10,000} = \dfrac{1}{5}$

9. $\Pr(\text{div. by 3 or 5}) = \Pr(\text{div. by 3}) + \Pr(\text{div. by 5}) - \Pr(\text{div. by 15}) = \dfrac{3333}{10,000} + \dfrac{1}{5} - \dfrac{666}{10,000} = .4667$

10. $\Pr(\text{div. by 3 or 12}) = \Pr(\text{div. by 3}) = \dfrac{3333}{10,000} = .3333$

11. $\dfrac{7}{8 + 7} = \dfrac{7}{15}$

12. **a.** $\dfrac{2 \cdot 1}{5 \cdot 4} = \dfrac{1}{10}$

   b. $4 \times \dfrac{1}{10} = \dfrac{2}{5}$

13. a.   0; at most 6 can appear.

    b. $\dfrac{C(2,\,2)C(8,\,4)}{C(10,\,6)} = \dfrac{1 \cdot 70}{210} = \dfrac{1}{3}$

14. a. $\left(\dfrac{1}{36}\right)^3$

    b. $\left(\dfrac{1}{10}\right)^4$

    c. $\dfrac{3}{10}$

    d. $\left(\dfrac{26}{36}\right)^3\left(\dfrac{1}{2}\right)^4 = \left(\dfrac{13}{18}\right)^3\left(\dfrac{1}{2}\right)^4$

15. a. $\dfrac{1}{4} \times \dfrac{1}{3} = \dfrac{1}{12}$

    b. $\dfrac{1}{4} + \dfrac{1}{3} - \dfrac{1}{12} = \dfrac{1}{12}$

16. $\dfrac{7}{7+5} = \dfrac{7}{12}$

17. $\dfrac{2}{7} \times \dfrac{1}{6} = \dfrac{1}{21}$

18. $\Pr(3 \text{ is drawn on 1st or 2nd draw}) = \dfrac{1}{3} + \left(\dfrac{1}{3}\right)\left(\dfrac{1}{2}\right) + \left(\dfrac{1}{3}\right)\left(\dfrac{1}{2}\right) = \dfrac{2}{3}$

19.   No; $\Pr\left(F|E\right) = \dfrac{1}{6} > \Pr(F) = \dfrac{5}{36}$

20. $\Pr\left(E|F\right) = \dfrac{\Pr(E \cap F)}{\Pr(F)} = \dfrac{\Pr(E) + \Pr(F) - \Pr(E \cup F)}{\Pr(F)} = \dfrac{.4 + .3 - .5}{.3} = \dfrac{.2}{.3} = \dfrac{2}{3}$

21. $\Pr\left(C|\text{wrong}\right) = \dfrac{\Pr(C) \times \Pr\left(\text{wrong}|C\right)}{\Pr(C) \times \Pr\left(\text{wrong}|C\right) + \Pr(A) \times \Pr\left(\text{wrong}|A\right) + \Pr(B) \times \Pr\left(\text{wrong}|B\right)}$

      $= \dfrac{.20 \times .05}{(.20 \times .05) + (.40 \times .02) + (.40 \times .03)} = \dfrac{.01}{.03} = \dfrac{1}{3}$

22. $1 - \left(1 \times \dfrac{6}{7} \times \dfrac{5}{7}\right) = \dfrac{19}{49}$

23. $\Pr(A \cup B) = \frac{1}{2}; \Pr(A' \cap B) = \frac{1}{3}$

$\Pr(A) + \Pr(A' \cap B) = \Pr(A \cup B)$

$\Pr(A) = \frac{1}{2} - \frac{1}{3} = \frac{1}{6}$

24. [number of ways to draw 7 balls] $= \frac{10!}{3!} = 604,800$

[number of ways to draw 3 odd-numbered balls or odd-numbered draws] $= C(4, 3) \cdot 5 \cdot 4 \cdot 3 \cdot 5 \cdot 4 \cdot 3 \cdot 2$
$= 28,800$

$\Pr(\text{draw 3 odd-numbered balls on odd-numbered draws}) = \frac{28,800}{604,800} = \frac{1}{21}$

25. $\Pr\left(\text{other is \$5}|\$5\right) = \frac{\Pr(\text{both are \$5})}{\Pr(\$5)} = \frac{\frac{1}{3}}{\frac{1}{2}} = \frac{2}{3}$

26. Let S = selected cup, M = moved unselected cup, U = unmoved unselected cup.

$\Pr\left(\text{under } S|\text{moved } M\right) = \dfrac{\Pr(\text{under } S) \cdot \Pr\left(\text{move } M|\text{under } S\right)}{\Pr(\text{under } S) \cdot \Pr\left(\text{moved } M|\text{under } S\right) + \Pr(\text{not under } S) \cdot \Pr\left(\text{moved } M|\text{not under } S\right)}$

$= \dfrac{\frac{1}{3} \cdot \frac{1}{2}}{\frac{1}{3} \cdot \frac{1}{2} + \frac{2}{3} \cdot \frac{1}{2}} = \frac{1}{3}$

$\Pr\left(\text{under } U|\text{moved } M\right) = 1 - \left[\Pr\left(\text{under } S|\text{moved } M\right) + \Pr\left(\text{under } M|\text{moved } M\right)\right] = 1 - \left(\frac{1}{3} + 0\right) = \frac{2}{3}$

You should switch.

27. $\dfrac{13}{13 + 12} = .52$

28. $.26\% = \dfrac{13}{50}$; $a = 13$ and $a + b = 50$, so $b = 37$. The odds are 13 to 37.

29. $\dfrac{3}{9!} = \dfrac{1}{120,960}$

30. If n is the number of dragons, then $\dfrac{\text{\# of heads on 1-headed dragons}}{\text{\# of heads}} = \dfrac{\frac{n}{3}}{\frac{n}{3} + 2 \cdot \frac{n}{3} + 3 \cdot \frac{n}{3}} = \dfrac{1}{6}$

31. $\Pr(\text{no tails}) + \Pr(\text{1 tail each}) + \Pr(\text{2 tails each}) + \Pr(\text{3 tails each}) = \left(\frac{1}{8}\right)^2 + \left(\frac{3}{8}\right)^2 + \left(\frac{3}{8}\right)^2 + \left(\frac{1}{8}\right)^2 = \frac{5}{16}$

32. $\Pr\left(\text{one 3}|\text{no doubles}\right) = \dfrac{5 + 5}{30} = \dfrac{1}{3}$

33. $\Pr\left(\text{both parents left - handed}\mid\text{child left - handed}\right) = \dfrac{\Pr(\text{all three left - handed})}{\Pr(\text{child left - handed})}$

$= \dfrac{.4 \times .25 \times .25}{(.4 \times .25 \times .25)+(.2 \times .25 \times .75)+(.2 \times .75 \times .25)+(.1 \times .75 \times .75)} = \dfrac{4}{25}$

34. $\dfrac{120 - (\text{speak Chinese or Spanish}) - (\text{speak French only})}{120} = \dfrac{120 - (30 + 50 - 12) - (75 - 30 - 15 + 7)}{120} = \dfrac{1}{8}$

35. There are 100 numbers (from 200–299) with a two in the hundreds digit.
There are 10 numbers (from 120–129) with a two in the tens digit.
There are 10 numbers (from 320–329) with a two in the tens digit.
There are 9 numbers (102, 112, 132, 142, …, 192) with a two in the ones digit. There are 9 numbers
(302, 312, 332, 342, …, 392) with a two in the ones digit.
The probability is $\dfrac{100 + 10 + 10 + 9 + 9}{301} = \dfrac{138}{301}$.

36. $\dfrac{w + 3}{(w + 3) + (m + 2)} = \dfrac{w + 3}{w + m + 5}$
(e)

37. $(.60 \times .90) + (.40 \times .05) = .56 = 56\%$
(a)

38. $1 - \dfrac{365 \times 364 \times 363 \times \cdots \times (365 - n + 1)}{365^n} > .80$ when $n \geq 35$.

39. n = number sick
Solve: $n \times .96 + (12{,}735 - n) \times .02 = 650$
$n \approx 421$

40. Use seq(int(6*rand)+1,X,1,15,1)

41. Use seq(int(52*rand)+1,X,1,3,1), with spades = 1 through 13.

Theoretical probability: $\Pr = \Pr(2 \text{ spades}) + \Pr(3 \text{ spades}) = 3\left(\dfrac{1}{4} \cdot \dfrac{1}{4} \cdot \dfrac{3}{4}\right) + \left(\dfrac{1}{4} \cdot \dfrac{1}{4} \cdot \dfrac{1}{4}\right) = \dfrac{5}{32}$

Chapter 7

Exercises 1

1.

Grade	Relative Frequency
0	$\frac{2}{25} = .08$
1	$\frac{3}{25} = .12$
2	$\frac{10}{25} = .40$
3	$\frac{6}{25} = .24$
4	$\frac{4}{25} = .16$

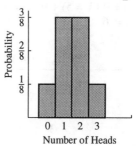

3.

Number of calls during minute	Relative Frequency
20	$\frac{3}{60} = .05$
21	$\frac{3}{60} = .05$
22	$\frac{0}{60} = 0$
23	$\frac{6}{60} = .10$
24	$\frac{18}{60} = .30$
25	$\frac{12}{60} = .20$
26	$\frac{0}{60} = 0$
27	$\frac{9}{60} = .15$
28	$\frac{6}{60} = .10$
29	$\frac{3}{60} = .05$

5. HHH, HHT, HTH, THH, HTT, THT, TTH, TTT

Number of Heads	Probability
0	$\dfrac{\binom{3}{0}}{2^3} = \dfrac{1}{8}$
1	$\dfrac{\binom{3}{1}}{2^3} = \dfrac{3}{8}$
2	$\dfrac{\binom{3}{2}}{2^3} = \dfrac{3}{8}$
3	$\dfrac{\binom{3}{3}}{2^3} = \dfrac{1}{8}$

7. Probability of hit $= \dfrac{1}{3}$

Probability of miss $= \dfrac{2}{3}$

Number of Shots	Probability
1	$\frac{1}{3}$
2	$\left(\frac{2}{3}\right)\left(\frac{1}{3}\right) = \frac{2}{9}$
3	$\left(\frac{2}{3}\right)^2\left(\frac{1}{3}\right) = \frac{4}{27}$
4	$\left(\frac{2}{3}\right)^3 = \frac{24}{81}$

9.

No. Red Balls	Player's Earnings	Probability
2	$5	$\dfrac{\binom{2}{2}\binom{4}{0}}{\binom{6}{2}} = \dfrac{1}{15}$
1	$1	$\dfrac{\binom{2}{1}\binom{4}{1}}{\binom{6}{2}} = \dfrac{8}{15}$
0	−1$	$\dfrac{\binom{2}{0}\binom{4}{2}}{\binom{6}{2}} = \dfrac{6}{15}$

11. $\Pr(5 \le X \le 7) = \Pr(X = 5) + \Pr(X = 6) + \Pr(X = 7)$
$= .2 + .1 + .3 = .6$

13.

k	$\Pr(X^2 = k)$
0	.1
1	.2
4	.3
9	.2
16	.2

15.

k	$\Pr(X - 1 = k)$
−1	.1
0	.2
1	.3
2	.2
3	.2

17.

k	$Pr\left(\frac{1}{5}Y = k\right)$
1	.3
2	.4
3	.1
4	.1
5	.1

19.

$(X+1)^2 = k$	$Pr\left((X+1)^2 = k\right)$
$(0+1)^2 = 1$	.1
$(1+1)^2 = 4$	.2
$(2+1)^2 = 9$	.3
$(3+1)^2 = 16$	.2
$(4+1)^2 = 25$	.2

21.

	Relative Frequency	
Grade	9 A.M. class	10 A.M. class
F	$\frac{10}{60} \approx .17$	$\frac{16}{100} = .16$
D	$\frac{15}{60} = .25$	$\frac{23}{100} = .23$
C	$\frac{20}{60} \approx .33$	$\frac{15}{100} = .15$
B	$\frac{10}{60} \approx .17$	$\frac{21}{100} = .21$
A	$\frac{5}{60} \approx .08$	$\frac{25}{100} = .25$

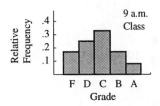

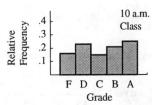

The 9 A.M. class has the distribution centered on the C grade with relatively few A's. The 10A.M. class has a large percentage of A's and D's with fewer C's.

23. percentage with C or higher:

$$\left(\frac{10+6+4}{25}\right) \times 100\% = 80\%$$

25. a. less than 22: $3 + 3 = 6$
more than 27: $6 + 3 = 9$

combined: $\left(\frac{6+9}{60}\right) \times 100\% = 25\%$

b. between 23 and 25:

$$\left(\frac{6+18+12}{60}\right) \times 100\% = 60\%$$

c.

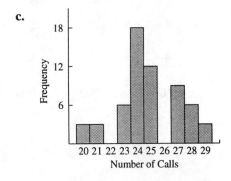

d. Estimated average number of calls would be 24 since that number has the highest frequency of occurrence. It is actually ≈ 25.

27. a. 59

    b. $\left(\dfrac{2}{40}\right) \times 100\% = 5\%$ of the time

    c. 54

    d. $\left(\dfrac{8+4+2}{40}\right) \times 100\% = 35\%$

    e. Estimate an average of 54 items produced.

29. a. $\Pr(U = 4) = 1 - \left(\dfrac{3}{15} + \dfrac{2}{15} + \dfrac{4}{15} + \dfrac{5}{15}\right)$

               $= 1 - \dfrac{14}{15} = \dfrac{1}{15} \approx .07$

    b. $\Pr(U \geq 2) = \Pr(U = 2) + \Pr(U = 3) + \Pr(U = 4)$

               $= \dfrac{4}{15} + \dfrac{5}{15} + \dfrac{1}{15} = \dfrac{10}{15} = \dfrac{2}{3} \approx .67$

    c. $\Pr(U \leq 3) = 1 - \Pr(U = 4) = 1 - \dfrac{1}{15} = \dfrac{14}{15} \approx .93$

    d.

$U + 2 = K$	$\Pr(U + 2 = K)$
$0 + 2 = 2$	$\frac{3}{15}$
$1 + 2 = 3$	$\frac{2}{15}$
$2 + 2 = 4$	$\frac{4}{15}$
$3 + 2 = 5$	$\frac{5}{15}$
$4 + 2 = 6$	$\frac{1}{15}$

        $\Pr(U + 2 < 4) = \Pr(U = 2) + \Pr(U = 3) = \dfrac{3}{15} + \dfrac{2}{15} = \dfrac{5}{15} = \dfrac{1}{3} \approx .33$

e.

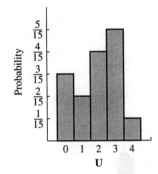

Exercises 2

1. Let "success" be the outcome "one." Then $p = \frac{1}{6}$, $q = \frac{5}{6}$, $n = 4$.

$$\Pr(X = 2) = \binom{4}{2}\left(\frac{1}{6}\right)^2\left(\frac{5}{6}\right)^2 = \frac{25}{216} \approx .1157$$

3. Let "success" be a "sale." Then $p = \frac{1}{4}$, $q = \frac{3}{4}$, $n = 4$.

$$\Pr(X = 3) = \binom{4}{3}\left(\frac{1}{4}\right)^3\left(\frac{3}{4}\right) = \frac{3}{64} \approx .0469$$

5. Let "success" be "a vote for" the candidate. Then $p = .6$, $q = .4$, $n = 5$.
   $\Pr(X \le 2) = \Pr(X = 0) + \Pr(X = 1) + \Pr(X = 2)$

   $$= \binom{5}{0}(.6)^0(.4)^5 + \binom{5}{1}(.6)^1(.4)^4 + \binom{5}{2}(.6)^2(.4)^3$$

   $$= .01024 + .0768 + .2304 \approx .3174$$

7. Let "success" be "choose a woman." Then $p = .6$, $q = .4$, $n = 5$.
   $\Pr(X \le 1) = \Pr(X = 0) + \Pr(X = 1)$

   $$= \binom{5}{0}(.6)^0(.4)^5 + \binom{5}{1}(.6)^1(.4)^4$$

   $$= .01024 + .0768 \approx .0870$$

   It is only on a rare occasion that no or only 1 woman would be selected for the committee. The selection may not have been random.

9. Let "success" be "a car with a commuter sticker." Then $p = .3$, $q = .7$, $n = 10$.
   $\Pr(X \ge 2) = 1 - \Pr(X < 2) = 1 - [\Pr(X = 0) + \Pr(X = 1)]$

   $$= 1 - \left[\binom{10}{0}(.3)^0(.7)^{10} + \binom{10}{1}(.3)^1(.7)^9\right]$$

   $$\approx 1 - (.0282 + .1211) = .8507$$

11. Let "success" be a "five." Then $p = \dfrac{1}{6}$, $q = \dfrac{5}{6}$, $n = 12$.

$\Pr(X \geq 2) = 1 - \Pr(X < 2)$
$= 1 - [\Pr(X = 0) + \Pr(X = 1)]$

$1 - \left[\binom{12}{0}\left(\dfrac{1}{6}\right)^0 \left(\dfrac{5}{6}\right)^{12} + \binom{12}{1}\left(\dfrac{1}{6}\right)^1 \left(\dfrac{5}{6}\right)^{11} \right]$

$\approx 1 - (.11216 + .26918) \approx .6187$

13. Let "success" be "brand X." Then $p = .2$,
$q = .8, n = 9$.
$\Pr(X > 2) = 1 - \Pr(X \leq 2)$
$= 1 - [\Pr(X = 0) + \Pr(X = 1) + \Pr(X = 2)]$

$= 1 - \left[\binom{9}{0}(.2)^0(.8)^9 + \binom{9}{1}(.2)^1(.8)^8 + \binom{9}{2}(.2)^2(.8)^7 \right] \approx 1 - (.1342 + .3020 + .3020) = .2618$

15.

k	$\Pr(X = k)$
0	$\binom{8}{0}(.4)^0(.6)^8 \approx .0168$
1	$\binom{8}{1}(.4)^1(.6)^7 \approx .0896$
2	$\binom{8}{2}(.4)^2(.6)^6 \approx .2090$
3	$\binom{8}{3}(.4)^3(.6)^5 \approx .2787$
4	$\binom{8}{4}(.4)^4(.6)^4 \approx .2322$
5	$\binom{8}{5}(.4)^5(.6)^3 \approx .1239$
6	$\binom{8}{6}(.4)^6(.6)^2 \approx .0413$
7	$\binom{8}{7}(.4)^7(.6)^1 \approx .0079$
8	$\binom{8}{8}(.4)^8(.6)^0 \approx .0007$

17. Let "success" be "a guilty vote."
Then $p = .80$, $q = .20$, $n = 12$.
$\Pr(X \geq 10) = \Pr(X = 10) + \Pr(X = 11) + \Pr(X = 12)$
$= \binom{12}{10}(.8)^{10}(.2)^2 + \binom{12}{11}(.8)^{11}(.2)^1 + \binom{12}{12}(.8)^{12}(.2)^0 \approx .5583$

19. Let "success" be "six." Then $p = \frac{1}{6}$, $q = \frac{5}{6}$, $n = 10$.

$\Pr(X = 9) = \binom{10}{9}\left(\frac{1}{6}\right)^9\left(\frac{5}{6}\right)^1 = \frac{50}{6^{10}} \approx .00000083$

$\Pr(X \geq 9) = \Pr(X = 9) + \Pr(X = 10)$

$= \binom{10}{9}\left(\frac{1}{6}\right)^9\left(\frac{5}{6}\right)^1 + \binom{10}{10}\left(\frac{1}{6}\right)^{10}\left(\frac{5}{6}\right)^0 = \frac{51}{6^{10}} \approx .00000084$

$\Pr(X = 9 | X \geq 9) = \dfrac{\Pr(X = 9)}{\Pr(X = 9) + \Pr(X = 10)} = \dfrac{50}{51} \approx .9804$

21. Let "success" be "defective." Then $p = .02$, $q = .98$, $n = 300$.

a. $\Pr(X = 12) = \binom{300}{12}(.02)^{12}(.98)^{288} \approx .0108$

b. $\Pr(X \leq 9) \approx .9182$

23. Let "success" be "chosen for jury duty." Then $p = .004$, $q = .996$, $n = 900$.

a. $\Pr(X = 8) = \binom{900}{8}(.004)^8(.996)^{892} \approx .0190$

b.

k	$\Pr(X = k)$
0	.0271
1	.0981
2	.1770
3	.2128
4	.1916
5	.1379
6	.0826

c. $\Pr(X \leq 9) \approx .9961$

Exercises 3

1. $E(X) = 0(.15) + 1(.2) + 2(.1) + 3(.25) + 4(.3) = 2.35$

3. a. $\text{GPA} = \dfrac{4+4+4+3+3+3+3+2+2+1}{10} = \dfrac{29}{10} = 2.9$

b.

Grade	Relative Frequency
4	$\frac{3}{10} = .3$
3	$\frac{4}{10} = .4$
2	$\frac{2}{10} = .2$
1	$\frac{1}{10} = .1$

c. $E(X) = 4(.3) + 3(.4) + 2(.2) + 1(.1) = 2.9$

5. $\bar{x}_A = 0(.3) + 1(.3) + 2(.2) + 3(.1) + 4(0) + 5(.1) = 1.5$
$\bar{x}_B = 0(.2) + 1(.3) + 2(.3) + 3(.1) + 4(.1) + 5(0) = 1.6$
Group *A* had fewer cavities.

7.

Earnings	Probability
–$1	$\frac{37}{38}$
$35	$\frac{1}{38}$

$E(X) = -1\left(\dfrac{37}{38}\right) + 35\left(\dfrac{1}{38}\right) \approx -.0526$

9.

Earnings	Probability
–50¢	$\frac{2}{6} = \frac{1}{3}$
0¢	$\left(\frac{4}{6}\right)\left(\frac{2}{5}\right) = \frac{4}{15}$
50¢	$\left(\frac{4}{6}\right)\left(\frac{3}{5}\right)\left(\frac{2}{4}\right) = \frac{1}{5}$
$1	$\left(\frac{4}{6}\right)\left(\frac{3}{5}\right)\left(\frac{2}{4}\right)\left(\frac{2}{3}\right) = \frac{2}{15}$
$1.50	$\left(\frac{4}{6}\right)\left(\frac{3}{5}\right)\left(\frac{2}{4}\right)\left(\frac{1}{3}\right)\left(\frac{2}{2}\right) = \frac{1}{15}$

$E(X) = -.5\left(\dfrac{1}{3}\right) + 0\left(\dfrac{4}{15}\right) + .5\left(\dfrac{1}{5}\right) + 1\left(\dfrac{2}{15}\right) + 1.5\left(\dfrac{1}{15}\right) \approx .1667$

11. Let *x* be the cost of the policy.
$\mu = (-x)(.9) + (10{,}000 - x)(.1) = -1.2x + 1000$
The expected value is zero if $x = 1000$.
He should be willing to pay up to $1000.

13.

Recorded Value	Probability
1	$\frac{1}{36}$
2	$\frac{3}{36}$
3	$\frac{5}{36}$
4	$\frac{7}{36}$
5	$\frac{9}{36}$
6	$\frac{11}{36}$

$$E(X) = 1\left(\frac{1}{36}\right) + 2\left(\frac{3}{36}\right) + 3\left(\frac{5}{36}\right) + 4\left(\frac{7}{36}\right) + 5\left(\frac{9}{36}\right) + 6\left(\frac{11}{36}\right) \approx 4.47$$

15. $\dfrac{240 + 80 + 110 + x}{4} = 150$

$x = 150 \cdot 4 - (240 + 80 + 110) = 170$

(b)

17. Cost for \$5 books is $5x$.
Cost for \$8 books is $8y$.
The total costs are $5x + 8y$.
The total number of books is $x + y$.

$\overline{X} = \dfrac{5x + 8y}{x + y}$ (a)

19. Let x be the first year income.

$\dfrac{x + 1.5x + 2.5x}{3} = 45,000$

$5x = 135,000$

$x = 27,000$

Second year income $= 1.5x = 1.5(27,000) = 40,500$ (d)

21. Let x be the greatest number of people.

$\left(\dfrac{1}{2}x\right) \cdot 180 + \left(\dfrac{1}{2}x\right) \cdot 215 \le 2000$

$197.5x \le 2000$

$x \le 10.13$ (d)

Exercises 4

1. $m = 70(.5) + 71(.2) + 72(.1) + 73(.2) = 71$

$\sigma^2 = (70 - 71)^2(.5) + (71 - 71)^2(.2) + (72 - 71)^2(.1) + (73 - 71)^2(.2)$

$= .5 + 0 + .1 + .8 = 1.4$

3. *B*

5. a. $\mu_A = -10\left(\dfrac{1}{5}\right) + 20\left(\dfrac{3}{5}\right) + 25\left(\dfrac{1}{5}\right) = 15$

$\mu_B = 0(.3) + 10(.4) + 30(.3) = 13$

$\sigma_A^2 = (-10-15)^2\left(\dfrac{1}{5}\right) + (20-15)^2\left(\dfrac{3}{5}\right) + (25-15)^2\left(\dfrac{1}{5}\right) = 125 + 15 + 20 = 160$

$\sigma_B^2 = (0-13)^2(.3) + (10-13)^2(.4) + (30-13)^2(.3) = 50.7 + 3.6 + 86.7 = 141$

   b. Investment *A*

   c. Investment *B*

7. a. $\mu_A = 100(.1) + 101(.2) + 102(.3) + 103(0) + 104(0) + 105(.2) + 106(.2) = 103$

$\sigma_A^2 = (100-103)^2(.1) + (101-103)^2(.2) + \cdots + (106-103)^2(.2) = 4.6$

$\mu_B = 100(0) + 101(.2) + 102(0) + 103(.2) + 104(.1) + 105(.2) + 106(.3) = 104$

$\sigma_A^2 = (100-104)^2(0) + (101-104)^2(.2) + \cdots + (106-104)^2(.3) = 3.4$

   b. Business *B*

   c. Business *B*

9. a. $35 - c = 25$ and $35 + c = 45 \Rightarrow c = 10$.

Probability $\geq 1 - \dfrac{5^2}{10^2} = 1 - \dfrac{25}{100} = .75$

   b. $35 - c = 20$ and $35 + c = 50 \Rightarrow c = 15$

Probability $\geq 1 - \dfrac{5^2}{15^2} \approx .89$

   c. $35 - c = 29$ and $35 + c = 41 \Rightarrow c = 6$

Probability $\geq 1 - \dfrac{5^2}{6^2} \approx .31$

11. $\mu = 3000,\ \sigma = 250$

$3000 - c = 2000$ and $3000 + c = 4000 \Rightarrow c = 1000$

Probability $\geq 1 - \dfrac{250^2}{1000^2} = .9375$

Number of bulbs to replace: $\geq 5000(.9375) \approx 4688$

13. Probability $= 1 - \dfrac{6^2}{c^2} = \dfrac{7}{16}$

$16c^2 - 576 = 7c^2$

$9c^2 = 576$

$c = \sqrt{64} = 8$

15. a. $\mu = 2\left(\dfrac{1}{36}\right) + 3\left(\dfrac{2}{36}\right) + \cdots + 12\left(\dfrac{1}{36}\right) = 7$

$\sigma^2 = (2-7)^2\left(\dfrac{1}{36}\right) + \cdots + (12-7)^2\left(\dfrac{1}{36}\right) = \dfrac{210}{36} = \dfrac{35}{6}$

b. $\Pr(4 \le X \le 10) = \dfrac{3}{36} + \dfrac{4}{36} + \dfrac{5}{36} + \dfrac{6}{36} + \dfrac{5}{36} + \dfrac{4}{36} + \dfrac{3}{36} = \dfrac{30}{36} = \dfrac{5}{6}$

c. $7 - c = 4$ and $7 + c = 10 \Rightarrow c = 3$

Probability $\ge 1 - \dfrac{\frac{35}{6}}{3^2} = \dfrac{19}{54} \approx .35$

17. $E(X^2) = 0^2(.1) + 1^2(.3) + 2^2(.5) + 3^2(.1) = 3.2$

$\sigma^2 = E(X^2) - \mu^2 = 3.2 - (1.6)^2 = .64$

19.

$2X$	Probability
-2	$\frac{1}{8}$
-1	$\frac{3}{8}$
0	$\frac{1}{8}$
1	$\frac{1}{8}$
2	$\frac{2}{8}$

$\mu = -2\left(\dfrac{1}{8}\right) - 1\left(\dfrac{3}{8}\right) + \cdots + 2\left(\dfrac{2}{8}\right) = 0$

$\sigma^2 = (-2-0)^2\left(\dfrac{1}{8}\right) + \cdots + (2-0)^2\left(\dfrac{2}{8}\right) = 2$

$\sigma^2_{2X} = 4\sigma^2_X = 4\left(\dfrac{1}{2}\right) = 2$

21. $\bar{x} = 44,203.9$, $s \approx 5018.59$

23. $\bar{x} = 4.72$, $s \approx 1.43$

Exercises 5

1. $A(1.25) = .8944$

3. $1 - A(.25) = 1 - .5987 = .4013$

5. $A(1.5) - A(.5) = .9332 - .6915 = .2417$

7. $A(-.5) + (1 - A(.5)) = .3085 + (1 - .6915) = .6170$

9. $\Pr(Z \geq z) = .0401$
   $A(z) = 1 - .0401 = .9599$
   $z = 1.75$

11. $\Pr(-z \leq Z \leq z) = .5468$
    $A(-z) = \dfrac{1 - .5468}{2} = .2266$
    $-z = -.75$
    $z = .75$

13. $\mu = 6, \sigma = 2$

15. $\mu = 9, \sigma = 1$

17. $\dfrac{6-8}{\frac{3}{4}} = -\dfrac{2}{1} \cdot \dfrac{4}{3} = -\dfrac{8}{3}$

19. $\dfrac{x-8}{\frac{3}{4}} = 10$

    $x = \dfrac{30}{4} + 8 = \dfrac{62}{4} = 15\dfrac{1}{2}$

21. $\Pr(X \geq 9) = \Pr\left(Z \geq \dfrac{9-10}{\frac{1}{2}}\right) = \Pr(Z \geq -2) = 1 - \Pr(Z \leq -2) = 1 - .0228 = .9772$

23. $\Pr(6 \leq X \leq 10) = \Pr\left(\dfrac{6-7}{2} \leq Z \leq \dfrac{10-7}{2}\right) = \Pr(-.50 \leq Z \leq 1.50) = .9332 - .3085 = .6247$

25. $\mu = 3.3, \sigma = .2$
    $\Pr(X \geq 4) = \Pr\left(Z \geq \dfrac{4-3.3}{.2}\right) = \Pr(Z \geq 3.5) = 1 - \Pr(Z \leq 3.5) = 1 - .9998 = .0002$

27. $\mu = 6, \sigma = .02$
    $\Pr(5.95 \leq X \leq 6.05) = \Pr\left(\dfrac{5.95-6}{.02} x \leq Z \leq \dfrac{6.05-6}{.02}\right) = \Pr(-2.5 \leq Z \leq 2.5) = .9938 - .0062 = .9876$

29. $\mu = 30{,}000$, $\sigma = 4000$

$$\Pr(X > 39{,}000) = \Pr\left(Z > \frac{39{,}000 - 30{,}000}{4000}\right) = \Pr(Z > 2.25) = 1 - \Pr(Z \leq 2.25)$$
$$= 1 - .9878 = .0122$$

31. $\mu = 520$, $\sigma = 75$

    a. $\Pr(Z \geq z) = .10 \Rightarrow z_{40} \approx 1.30$

$$\frac{x - \mu}{\sigma} = \frac{x - 520}{75} = 1.30$$
$$x_{90} = 1.30(75) + 520 = 617.5 \approx 618$$

    b. $\Pr(-z \leq Z \leq z) = .90$
       $\Pr(Z \leq -z) = .05 \Rightarrow z_{05} \approx -1.65$

$$\frac{x - \mu}{\sigma} = \frac{x - 520}{75} = -1.65 \Rightarrow x_{05} = 396.25 \approx 396$$
$$\frac{x - \mu}{\sigma} = \frac{x - 520}{75} = 1.65 \Rightarrow x_{95} = 643.75 \approx 644$$

       Between 396 and 644

33. $\mu = 30{,}000$, $\sigma = 5000$
    $\Pr(Z \leq z) = .02 \Rightarrow z_{02} \approx -2.05$

$$\frac{x - \mu}{\sigma} = \frac{x - 30{,}000}{5000} = -2.05 \Rightarrow x_{02} = 19{,}750$$
19,750 miles

35. $\mu = ?$, $\sigma = .25$

    a. $\Pr(Z > z) = .005 \Rightarrow z_{99.5} \approx 2.60$

$$\frac{x - \mu}{\sigma} = \frac{6 - \mu}{.25} = 2.60 \Rightarrow \mu \approx 5.35 \text{ ounces}$$

    b. $\Pr(Z > z) = .99 \Rightarrow z_{01} \approx -2.35$

$$\frac{x - \mu}{\sigma} = \frac{x - 5.35}{.25} = -2.35 \Rightarrow x_{01} \approx 4.76 \text{ ounces}$$

37.

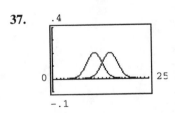

The curve is translated to the right.

39. $\mu = 5.4$, $\sigma = .6$

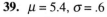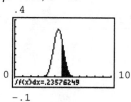

$\Pr(X > 5.832) \approx .2358$

Exercises 6

1. $n = 25$, $p = \dfrac{1}{5}$

$$\mu = np = 25\left(\frac{1}{5}\right) = 5, \quad \sigma = \sqrt{npq} = \sqrt{25\left(\frac{1}{5}\right)\left(\frac{4}{5}\right)} = 2$$

   a. $\Pr(X = 5) \approx \Pr\left(\dfrac{4.5 - 5}{2} \leq Z \leq \dfrac{5.5 - 5}{2}\right) = \Pr(-.25 \leq Z \leq .25) = .5987 - .4013 = .1974$

   b. $\Pr(3 \leq X \leq 7) \approx \Pr\left(\dfrac{2.5 - 5}{2} \leq Z \leq \dfrac{7.5 - 5}{2}\right) = \Pr(-1.25 \leq Z \leq 1.25) = .8944 - .1056 = .7888$

   c. $\Pr(X < 10) \approx \Pr\left(Z \leq \dfrac{9.5 - 5}{2}\right) = \Pr(Z \leq 2.25) = .9878$

3. $n = 20$, $p = \dfrac{1}{6}$

$$\mu = 20\left(\frac{1}{6}\right) = \frac{10}{3}, \quad \sigma = \sqrt{20\left(\frac{1}{6}\right)\left(\frac{5}{6}\right)} = \frac{5}{3}$$

$$\Pr(X \geq 8) \approx \Pr\left(Z \geq \dfrac{7.5 - \frac{10}{3}}{\frac{5}{3}}\right) = \Pr(Z \geq 2.5) = 1 - .9938 = .0062$$

5. $n = 100$, $p = \dfrac{1}{2}$

$$\mu = 100\left(\frac{1}{2}\right) = 50, \quad \sigma = \sqrt{100\left(\frac{1}{2}\right)\left(\frac{1}{2}\right)} = 5$$

$$\Pr(X \geq 63) \approx \Pr\left(Z \geq \dfrac{62.5 - 50}{5}\right) = \Pr(Z \geq 2.5) = 1 - .9938 = .0062$$

7. $n = 75, \ p = \dfrac{3}{4}$

$$\mu = 75\left(\dfrac{3}{4}\right) = 56.25, \ \sigma = \sqrt{75\left(\dfrac{3}{4}\right)\left(\dfrac{1}{4}\right)} = 3.75$$

$$\Pr(X \geq 68) \approx \Pr\left(Z \geq \dfrac{67.5 - 56.25}{3.75}\right) = \Pr(Z \geq 3) = 1 - .9987 = .0013$$

9. $n = 20, p = .310$

$$\mu = 20(.310) = 6.2, \ \sigma = \sqrt{20(.31)(.69)} \approx 2.068$$

$$\Pr(X \geq 6) \approx \Pr\left(Z \geq \dfrac{5.5 - 6.2}{2.068}\right) \approx \Pr(Z \geq -.34) \approx \Pr(Z \geq -.35) = 1 - .3632 = .6368$$

11. $n = 1000, p = .02$

$$\mu = 1000(.02) = 20, \ \sigma = \sqrt{1000(.02)(.98)} \approx 4.427$$

$$\Pr(X < 15) \approx \Pr\left(Z \leq \dfrac{14.5 - 20}{4.427}\right) \approx \Pr(Z \leq -1.24) \approx \Pr(Z \leq -1.25) = .1056$$

13. Estimated probability: $\dfrac{50}{250} = .2$

$n = 250, p = .25$

$$\mu = 250(.25) = 62.5, \ \sigma = \sqrt{250(.25)(.75)} \approx 6.847$$

$$\Pr(X \leq 50) \approx \Pr\left(Z \leq \dfrac{50.5 - 62.5}{6.847}\right) \approx \Pr(Z \leq -1.75) = .0401$$

15. $n = 100, \ p = \dfrac{1}{2}$

Exact: $\Pr(49 \leq X \leq 51) \approx .2356$

Normal Approximation: $\mu = 100\left(\dfrac{1}{2}\right) = 50, \ \sigma = \sqrt{100\left(\dfrac{1}{2}\right)\left(\dfrac{1}{2}\right)} = 5$

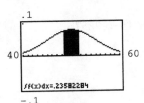

$\Pr(48.5 \leq X \leq 51.5) \approx .2358$

17. $n = 150, p = .2$

Exact: $\Pr(X = 30) \approx .0812$

Normal Approximation: $\mu = 150(.2) = 30, \ \sigma = \sqrt{150(.2)(.8)} \approx 4.899$

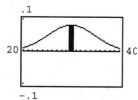

$\Pr(29.5 \le X \le 30.5) \approx .0813$

Chapter 7 Supplementary Exercises

1. $n = 3$, $p = \dfrac{1}{3}$

a.

k	$\Pr(X = k)$
0	$\binom{3}{0}\left(\frac{1}{3}\right)^0\left(\frac{2}{3}\right)^3 = \frac{8}{27}$
1	$\binom{3}{1}\left(\frac{1}{3}\right)^1\left(\frac{2}{3}\right)^2 = \frac{12}{27}$
2	$\binom{3}{2}\left(\frac{1}{3}\right)^2\left(\frac{2}{3}\right)^1 = \frac{6}{27}$
3	$\binom{3}{3}\left(\frac{1}{3}\right)^3\left(\frac{2}{3}\right)^0 = \frac{1}{27}$

b. $\mu = 0\left(\dfrac{8}{27}\right) + 1\left(\dfrac{12}{27}\right) + 2\left(\dfrac{6}{27}\right) + 3\left(\dfrac{1}{27}\right) = 1$

$\sigma^2 = (0-1)^2\left(\dfrac{8}{27}\right) + (1-1)^2\left(\dfrac{12}{27}\right) + (2-1)^2\left(\dfrac{6}{27}\right) + (3-1)^2\left(\dfrac{1}{27}\right) = \dfrac{2}{3}$

2. $\Pr(Z \ge .75) = 1 - .7734 = .2266$

3. $\Pr(6.5 \le X \le 11) = \Pr\left(\dfrac{6.5-5}{3} \le Z \le \dfrac{11-5}{3}\right) = A(2) - A(.5) = .9772 - .6915 = .2857$

4. $n = 4$, $p = .3$

$\Pr(X = 2) = \binom{4}{2}(.3)^2(.7)^2 = .2646$

5. $\mu = 10$, $\sigma = \dfrac{1}{3}$

$10 - c = 9$ and $10 + c = 11 \Rightarrow c = 1$

Probability: $\ge 1 - \dfrac{\left(\frac{1}{3}\right)^2}{1^2} = \dfrac{8}{9} \approx .89$

6. $\mu = 0(.2) + 1(.3) + 5(.1) + 10(.4) = 4.8$

$\sigma^2 = (0 - 4.8)^2(.2) + (1 - 4.8)^2(.3) + (5 - 4.8)^2(.1) + (10 - 4.8)^2(.4) = 19.76$

7. $\mu = 5.75,\ \sigma = .2$

$\Pr(X \geq 6) = \Pr\left(Z \geq \dfrac{6 - 5.75}{.2}\right) = \Pr(Z \geq 1.25) = 1 - .8944 = .1056$

10.56%

8. Let x be the number of red balls.

k	$\Pr(X = k)$
0	$\dfrac{\binom{4}{0}\binom{4}{4}}{\binom{8}{4}} = \dfrac{1}{70}$
1	$\dfrac{\binom{4}{1}\binom{4}{3}}{\binom{8}{4}} = \dfrac{16}{70}$
2	$\dfrac{\binom{4}{2}\binom{4}{2}}{\binom{8}{4}} = \dfrac{36}{70}$
3	$\dfrac{\binom{4}{3}\binom{4}{1}}{\binom{8}{4}} = \dfrac{16}{70}$
4	$\dfrac{\binom{4}{4}\binom{4}{0}}{\binom{8}{4}} = \dfrac{1}{70}$

$\mu = 0\left(\dfrac{1}{70}\right) + 1\left(\dfrac{16}{70}\right) + 2\left(\dfrac{36}{70}\right) + 3\left(\dfrac{16}{70}\right) + 4\left(\dfrac{1}{70}\right) = 2$

$\sigma^2 = (0 - 2)^2\left(\dfrac{1}{70}\right) + (1 - 2)^2\left(\dfrac{16}{70}\right) + (2 - 2)^2\left(\dfrac{36}{70}\right) + (3 - 2)^2\left(\dfrac{16}{70}\right) + (4 - 2)^2\left(\dfrac{1}{70}\right) = \dfrac{4}{7}$

9. $n = 54,\ p = \dfrac{2}{5}$

$\mu = 54\left(\dfrac{2}{5}\right) = 21.6$

$$\sigma = \sqrt{54\left(\frac{2}{5}\right)\left(\frac{3}{5}\right)} = 3.6$$

$$Pr(X \le 13) \approx Pr\left(Z \le \frac{13.5 - 21.6}{3.6}\right) = Pr(Z \le -2.25) = .0122$$

10. $n = 75, \ p = \frac{1}{4}$

$$\mu = 75\left(\frac{1}{4}\right) = 18.75$$

$$\sigma = \sqrt{75\left(\frac{1}{4}\right)\left(\frac{3}{4}\right)} = 3.75$$

$$Pr(8 \le X \le 22) \approx Pr\left(\frac{7.5 - 18.75}{3.75} \le Z \le \frac{22.5 - 18.75}{3.75}\right) = Pr(-3 \le Z \le 1) = .8413 - .0013 = .84$$

11. $\mu = 80, \ \sigma = 15$

$Pr(80 - n \le X \le 80 + h) = .8664$

$\dfrac{1 - .8664}{2} = .0668 \Rightarrow$ (area left of $80 - h$)

$Pr(Z \le z) = .0668$ when $z = -1.5$

$Pr(-1.5 \le Z \le 1.5) = .8664$

Therefore, $\dfrac{x - \mu}{\sigma} = -1.5$ and $\dfrac{x + \mu}{\sigma} = 1.5$.

$\dfrac{(80 - h) - 80}{15} = -1.5$ and $\dfrac{(80 + h) - 80}{15} = 1.5$

$h = 22.5$

12. $Pr(Z \ge z) = .7734$

$Pr(Z < z) = 1 - .7734 = .2266$

$z = -.75$

13. If exactly 4 games are played, one team has won 4 games in a row (all wins prior to the fourth and final game in $\binom{3}{3}$ ways) with probability $\left(\frac{1}{2}\right)^4$. If exactly 5 games are played, the winning team has three wins prior to the fifth and final game in $\binom{4}{3}$ ways with probability $\binom{4}{1}\left(\frac{1}{2}\right)^5$. The probability for either team winning the series in 5 games is $2 \cdot \binom{4}{3}\left(\frac{1}{2}\right)^5$. Continuing with this pattern, an n game series has probability $2 \cdot \binom{n-1}{3}\left(\frac{1}{2}\right)^n$.

n	$\Pr(X = n)$
4	$2 \cdot \binom{3}{3}\left(\frac{1}{2}\right)^4 = \frac{2}{16} = \frac{1}{8}$
5	$2 \cdot \binom{4}{3}\left(\frac{1}{2}\right)^5 = \frac{4}{16} = \frac{1}{4}$
6	$2 \cdot \binom{5}{3}\left(\frac{1}{2}\right)^6 = \frac{5}{16}$
7	$2 \cdot \binom{6}{3}\left(\frac{1}{2}\right)^7 = \frac{5}{16}$

14. There are 4 possible situations for each question. The table below shows the probability for each situation. That is, the student will guess true for a true question $(.6)(.6) = 36\%$ of the time and guess false for a true question $(.4)(.6) = 24\%$ of the time.

		Question	
		T	F
Guess	T	.36	.24
	F	.24	.16

The probability of a correct guess is $.36 + .16 = .52$ and the probability for an incorrect guess is $.24 + .24 = .48$. Therefore, the student's expected score will be:
(Probability of a correct guess) $\times$ (number of questions) $\times$ (points per question) $= (.52)(10)(10) = 52$
A better strategy is to choose true for all the questions for a score of 60 (or study the material so less guessing is involved).

Chapter 8

Exercises 1

1. Yes; the matrix is square, all entries are ≥ 0, and the sum of the entries in each column is 1.

3. No; the matrix is not square.

5. Yes; the matrix is square, all entries are ≥ 0, and the sum of the entries in each column is 1.

7. $\begin{bmatrix} \\ \end{bmatrix}_0 = \begin{bmatrix} .45 \\ .55 \end{bmatrix}_0$

$\begin{bmatrix} \\ \end{bmatrix}_1 = A\begin{bmatrix} \\ \end{bmatrix}_0 = \begin{bmatrix} .8 & .3 \\ .2 & .7 \end{bmatrix}\begin{bmatrix} .45 \\ .55 \end{bmatrix}_0$

$= \begin{bmatrix} .525 \\ .475 \end{bmatrix}_1$

$\begin{bmatrix} \\ \end{bmatrix}_2 = A^2\begin{bmatrix} \\ \end{bmatrix}_0 = \begin{bmatrix} .70 & .45 \\ .30 & .55 \end{bmatrix}\begin{bmatrix} .45 \\ .55 \end{bmatrix}_0$

$= \begin{bmatrix} .5625 \\ .4375 \end{bmatrix}_2$

After one generation, about 53% of French women will work. After two generations, about 56% of French women will work.

9. $\begin{bmatrix} \\ \\ \end{bmatrix}_0 = \begin{bmatrix} .40 \\ .40 \\ .20 \end{bmatrix}$

$\begin{bmatrix} \\ \\ \end{bmatrix}_1 = A\begin{bmatrix} \\ \\ \end{bmatrix}_0 = \begin{bmatrix} .5 & .4 & .2 \\ .4 & .3 & .6 \\ .1 & .3 & .2 \end{bmatrix}\begin{bmatrix} .4 \\ .4 \\ .2 \end{bmatrix}$

$= \begin{bmatrix} .4 \\ .4 \\ .2 \end{bmatrix}$

40% will be Zone I, 40% will be in Zone II, 20% will be in Zone III.

11. a. $\begin{array}{c} \\ S \\ M \\ L \end{array}\begin{array}{c} S \quad M \quad L \\ \begin{bmatrix} .4 & .5 & .3 \\ .6 & 0 & .2 \\ 0 & .5 & .5 \end{bmatrix} \end{array}$

b. $\begin{bmatrix} \\ \\ \end{bmatrix}_0 = \begin{bmatrix} .8 \\ .1 \\ .1 \end{bmatrix}_0$

$\begin{bmatrix} \\ \\ \end{bmatrix}_2 = A^2\begin{bmatrix} \\ \\ \end{bmatrix}_0$

$= \begin{bmatrix} .4 & .5 & .3 \\ .6 & 0 & .2 \\ 0 & .5 & .5 \end{bmatrix}^2\begin{bmatrix} .8 \\ .1 \\ .1 \end{bmatrix}_0 = \begin{bmatrix} .44 \\ .26 \\ .3 \end{bmatrix}_2$

44% will have a strenuous workout on Wednesday.

13. a. $\begin{array}{c} \\ D \\ R \end{array}\begin{array}{c} D \quad R \\ \begin{bmatrix} .7 & .4 \\ .3 & .6 \end{bmatrix} \end{array}$

b. $A^2 = \begin{bmatrix} .7 & .4 \\ .3 & .6 \end{bmatrix}\begin{bmatrix} .7 & .4 \\ .3 & .6 \end{bmatrix} = \begin{bmatrix} .61 & .52 \\ .39 & .48 \end{bmatrix}$

$A^3 = \begin{bmatrix} .7 & .4 \\ .3 & .6 \end{bmatrix}\begin{bmatrix} .61 & .52 \\ .39 & .48 \end{bmatrix}$

$= \begin{bmatrix} .583 & .556 \\ .417 & .444 \end{bmatrix}$

c. $\begin{bmatrix} \\ \end{bmatrix}_0 = \begin{bmatrix} 1 \\ 0 \end{bmatrix}_0$

$\begin{bmatrix} \\ \end{bmatrix}_3 = A^3\begin{bmatrix} \\ \end{bmatrix}_0$

$= \begin{bmatrix} .583 & .556 \\ .417 & .444 \end{bmatrix}\begin{bmatrix} 1 \\ 0 \end{bmatrix}_0 = \begin{bmatrix} .583 \\ .417 \end{bmatrix}_3$

58.3% will be Democrats.

15. a.
$$\begin{array}{c} & \begin{array}{ccc} U & S & R \end{array} \\ \begin{array}{c} U \\ S \\ R \end{array} & \left[\begin{array}{ccc} .86 & .05 & .03 \\ .08 & .86 & .05 \\ .06 & .09 & .92 \end{array}\right] \end{array}$$

b. Since we are concerned with only the people who live in urban areas in 1997, we use $\left[\ \right]_0 = \begin{bmatrix} 1 \\ 0 \\ 0 \end{bmatrix}_0$.

$$\left[\ \right]_2 = A^2 \left[\ \right]_0$$

$$= \begin{bmatrix} .86 & .05 & .03 \\ .08 & .86 & .05 \\ .06 & .09 & .92 \end{bmatrix} \begin{bmatrix} .86 & .05 & .03 \\ .08 & .86 & .05 \\ .06 & .09 & .92 \end{bmatrix} \begin{bmatrix} 1 \\ 0 \\ 0 \end{bmatrix}_0 = \begin{bmatrix} .7454 \\ .1406 \\ \mathbf{.114} \end{bmatrix}_2$$

11.4% of people who live in urban areas in 1997 will live in rural areas in 1999.

17. $A^1 = A = \begin{bmatrix} \frac{1}{3} & \frac{1}{3} \\ \frac{2}{3} & \frac{2}{3} \end{bmatrix} \approx \begin{bmatrix} .33 & .33 \\ .67 & .67 \end{bmatrix}$

$A^2 = A \cdot A = \begin{bmatrix} \frac{1}{3} & \frac{1}{3} \\ \frac{2}{3} & \frac{2}{3} \end{bmatrix} \begin{bmatrix} \frac{1}{3} & \frac{1}{3} \\ \frac{2}{3} & \frac{2}{3} \end{bmatrix}$

$= \begin{bmatrix} \frac{1}{3} & \frac{1}{3} \\ \frac{2}{3} & \frac{2}{3} \end{bmatrix} \approx \begin{bmatrix} .33 & .33 \\ .67 & .67 \end{bmatrix}$

The pattern continues. All powers are

approximately $\begin{bmatrix} .33 & .33 \\ .67 & .67 \end{bmatrix}$.

19. $A^1 = A = \begin{bmatrix} .1 & .3 \\ .9 & .7 \end{bmatrix}$

$A^2 = A \cdot A = \begin{bmatrix} .1 & .3 \\ .9 & .7 \end{bmatrix} \begin{bmatrix} .1 & .3 \\ .9 & .7 \end{bmatrix}$

$= \begin{bmatrix} .28 & .24 \\ .72 & .76 \end{bmatrix}$

$A^3 = A^2 \cdot A = \begin{bmatrix} .28 & .24 \\ .72 & .76 \end{bmatrix} \begin{bmatrix} .1 & .3 \\ .9 & .7 \end{bmatrix}$

$= \begin{bmatrix} .244 & .252 \\ .756 & .748 \end{bmatrix} \approx \begin{bmatrix} .24 & .25 \\ .76 & .75 \end{bmatrix}$

$A^4 = A^3 \cdot A = \begin{bmatrix} .244 & .252 \\ .756 & .748 \end{bmatrix} \begin{bmatrix} .1 & .3 \\ .9 & .7 \end{bmatrix}$

$= \begin{bmatrix} .2512 & .2496 \\ .7488 & .7504 \end{bmatrix} \approx \begin{bmatrix} .25 & .25 \\ .75 & .75 \end{bmatrix}$

$A^5 = A^4 \cdot A = \begin{bmatrix} .2512 & .2496 \\ .7488 & .7504 \end{bmatrix} \begin{bmatrix} .1 & .3 \\ .9 & .7 \end{bmatrix}$

$= \begin{bmatrix} .24976 & .25008 \\ .75024 & .74992 \end{bmatrix} \approx \begin{bmatrix} .25 & .25 \\ .75 & .75 \end{bmatrix}$

21. $A^1 = A = \begin{bmatrix} .2 & .2 & .2 \\ .3 & .3 & .3 \\ .5 & .5 & .5 \end{bmatrix}$

$A^2 = A \cdot A = \begin{bmatrix} .2 & .2 & .2 \\ .3 & .3 & .3 \\ .5 & .5 & .5 \end{bmatrix} \begin{bmatrix} .2 & .2 & .2 \\ .3 & .3 & .3 \\ .5 & .5 & .5 \end{bmatrix}$

$= \begin{bmatrix} .2 & .2 & .2 \\ .3 & .3 & .3 \\ .5 & .5 & .5 \end{bmatrix}$

The pattern continues. All powers are

$\begin{bmatrix} .2 & .2 & .2 \\ .3 & .3 & .3 \\ .5 & .5 & .5 \end{bmatrix}$.

23. No; all powers have a zero entry in the upper right corner, so the matrix is not regular.

25. a. Use the method described in the text to generate the next four distribution matrices.

$$\begin{bmatrix} .35 \\ .65 \end{bmatrix}, \begin{bmatrix} .425 \\ .575 \end{bmatrix}, \begin{bmatrix} .3875 \\ .6125 \end{bmatrix}, \begin{bmatrix} .40625 \\ .59375 \end{bmatrix}$$

b. $A^4 B = \begin{bmatrix} .40625 \\ .59375 \end{bmatrix}$

27. The distribution matrix gets closer and closer to $\begin{bmatrix} .4 \\ .6 \end{bmatrix}$.

29. The matrices get closer and closer to $\begin{bmatrix} .4 & .4 \\ .6 & .6 \end{bmatrix}$. Each column of this matrix is the same as the 2×1 matrix found in Exercise 27.

Exercises 2

1. Yes; the matrix is stochastic, and all entries are positive.

3. Yes; the matrix is stochastic, and the second power is $\begin{bmatrix} .79 & .3 \\ .21 & .7 \end{bmatrix}$, which has all positive entries.

5. Yes; the matrix is stochastic, and the second power is $\begin{bmatrix} .8 & .08 & .4 \\ .1 & .86 & .3 \\ .1 & .06 & .3 \end{bmatrix}$, which has all positive entries.

7. $\begin{cases} x + y = 1 \\ \begin{bmatrix} .5 & .1 \\ .5 & .9 \end{bmatrix}\begin{bmatrix} x \\ y \end{bmatrix} = \begin{bmatrix} x \\ y \end{bmatrix} \end{cases}$

$$\begin{cases} x + y = 1 \\ .5x + .1y = x \\ .5x + .9y = y \end{cases}$$

$$\begin{cases} x + y = 1 \\ -.5x + .1y = 0 \\ .5x - .1y = 0 \end{cases}$$

The second and third equations in this system are equivalent.

$$\begin{bmatrix} 1 & 1 & | & 1 \\ -.5 & .1 & | & 0 \end{bmatrix} \xrightarrow{[2]+.5[1]} \begin{bmatrix} 1 & 1 & | & 1 \\ 0 & .6 & | & .5 \end{bmatrix}$$

$$\xrightarrow{\frac{5}{3}[2]} \begin{bmatrix} 1 & 1 & | & 1 \\ 0 & 1 & | & \frac{5}{6} \end{bmatrix}$$

$$\xrightarrow{[1]+(-1)[2]} \begin{bmatrix} 1 & 0 & | & \frac{1}{6} \\ 0 & 1 & | & \frac{5}{6} \end{bmatrix}$$

$$x = \frac{1}{6}, \; y = \frac{5}{6}$$

The stable distribution is $\begin{bmatrix} x \\ y \end{bmatrix} = \begin{bmatrix} \frac{1}{6} \\ \frac{5}{6} \end{bmatrix}$.

9. $\begin{cases} x + y = 1 \\ \begin{bmatrix} .8 & .3 \\ .2 & .7 \end{bmatrix}\begin{bmatrix} x \\ y \end{bmatrix} = \begin{bmatrix} x \\ y \end{bmatrix} \end{cases}$

$$\begin{cases} x + y = 1 \\ .8x + .3y = x \\ .2x + .7y = y \end{cases}$$

$$\begin{cases} x + y = 1 \\ -.2x + .3y = 0 \\ .2x - .3y = 0 \end{cases}$$

The second and third equations in this system are equivalent.

$$\begin{bmatrix} 1 & 1 & | & 1 \\ -.2 & .3 & | & 0 \end{bmatrix} \xrightarrow{[2]+.2[1]} \begin{bmatrix} 1 & 1 & | & 1 \\ 0 & .5 & | & .2 \end{bmatrix}$$

$$\xrightarrow{2[2]} \begin{bmatrix} 1 & 1 & | & 1 \\ 0 & 1 & | & .4 \end{bmatrix}$$

$$\xrightarrow{[1]+(-1)[2]} \begin{bmatrix} 1 & 0 & | & .6 \\ 0 & 1 & | & .4 \end{bmatrix}$$

$$x = .6, \; y = .4$$

The stable distribution is $\begin{bmatrix} x \\ y \end{bmatrix} = \begin{bmatrix} .6 \\ .4 \end{bmatrix}$.

11.
$$\begin{cases} x+y+z=1 \\ \begin{bmatrix} .1 & .4 & .7 \\ .6 & .4 & .2 \\ .3 & .2 & .1 \end{bmatrix}\begin{bmatrix} x \\ y \\ z \end{bmatrix}=\begin{bmatrix} x \\ y \\ z \end{bmatrix} \end{cases}$$

$$\begin{cases} x + y + z = 1 \\ .1x + .4y + .7z = x \\ .6x + .4y + .2z = y \\ .3x + .2y + .1z = 2 \end{cases}$$

$$\begin{cases} x + y + z = 1 \\ -.9x + .4y + .7z = 0 \\ .6x - .6y + .2z = 0 \\ .3x + .2y - .9z = 0 \end{cases}$$

$$\begin{bmatrix} 1 & 1 & 1 & | & 1 \\ -.9 & .4 & .7 & | & 0 \\ .6 & -.6 & .2 & | & 0 \\ .3 & .2 & -.9 & | & 0 \end{bmatrix} \begin{matrix} \\ 10[2] \\ 10[3] \\ 10[4] \end{matrix} \longrightarrow \begin{bmatrix} 1 & 1 & 1 & | & 1 \\ -9 & 4 & \cdot & 7 & | & 0 \\ 6 & -6 & 2 & | & 0 \\ 3 & 2 & -9 & | & 0 \end{bmatrix}$$

$$\begin{matrix} [2]+9[1] \\ [3]+(-6)[1] \\ [4]+(-3)[1] \end{matrix} \longrightarrow \begin{bmatrix} 1 & 1 & 1 & | & 1 \\ 0 & 13 & 16 & | & 9 \\ 0 & -12 & -4 & | & -6 \\ 0 & -1 & -12 & | & -3 \end{bmatrix}$$

$$\begin{matrix} \frac{1}{13}[2] \\ [1]+(-1)[2] \\ [3]+12[2] \\ [4]+1[2] \end{matrix} \longrightarrow \begin{bmatrix} 1 & 0 & -\frac{3}{13} & | & \frac{4}{13} \\ 0 & 1 & \frac{16}{13} & | & \frac{9}{13} \\ 0 & 0 & \frac{140}{13} & | & \frac{30}{13} \\ 0 & 0 & -\frac{140}{13} & | & -\frac{30}{13} \end{bmatrix}$$

$$\begin{matrix} \frac{13}{140}[3] \\ [1]+\frac{3}{13}[3] \\ [2]-\frac{16}{13}[3] \\ [4]+\frac{140}{13}[3] \end{matrix} \longrightarrow \begin{bmatrix} 1 & 0 & 0 & | & \frac{5}{14} \\ 0 & 1 & 0 & | & \frac{3}{7} \\ 0 & 0 & 1 & | & \frac{3}{14} \\ 0 & 0 & 0 & | & 0 \end{bmatrix}$$

$$x=\frac{5}{14},\ y=\frac{3}{7},\ z=\frac{3}{14}$$

The stable distribution is $\begin{bmatrix} x \\ y \\ z \end{bmatrix}=\begin{bmatrix} \frac{5}{14} \\ \frac{3}{7} \\ \frac{3}{14} \end{bmatrix}$.

$$\begin{array}{c} & \begin{array}{ccc} S & M & L \end{array} \\ \begin{array}{c} S \\ M \\ L \end{array} & \begin{bmatrix} .4 & .5 & .3 \\ .6 & 0 & .2 \\ 0 & .5 & .5 \end{bmatrix} \end{array}$$

13. The stochastic matrix is (above).

Find the stable distribution.

$$\begin{cases} x + y + z = 1 \\ \begin{bmatrix} .4 & .5 & .3 \\ .6 & 0 & .2 \\ 0 & .5 & .5 \end{bmatrix}\begin{bmatrix} x \\ y \\ z \end{bmatrix} = \begin{bmatrix} x \\ y \\ z \end{bmatrix} \end{cases}$$

$$\begin{cases} x + y + z = 1 \\ .4x + .5y + .3z = x \\ .6x + .2z = y \\ .5y + .5z = z \end{cases}$$

$$\begin{cases} x + y + z = 1 \\ -.6x + .5y + .3z = 0 \\ .6x - y + .2z = 0 \\ .5y - .5z = 0 \end{cases}$$

$$\begin{bmatrix} 1 & 1 & 1 & | & 1 \\ -.6 & .5 & .3 & | & 0 \\ .6 & -1 & .2 & | & 0 \\ 0 & .5 & -.5 & | & 0 \end{bmatrix} \xrightarrow[\substack{10[2] \\ 10[3] \\ 2[4]}]{} \begin{bmatrix} 1 & 1 & 1 & | & 1 \\ -6 & 5 & 3 & | & 0 \\ 6 & -10 & 2 & | & 0 \\ 0 & 1 & -1 & | & 0 \end{bmatrix}$$

$$\xrightarrow[\substack{[2]+6[1] \\ [3]+(-6)[1]}]{} \begin{bmatrix} 1 & 1 & 1 & | & 1 \\ 0 & 11 & 9 & | & 6 \\ 0 & -16 & -4 & | & -6 \\ 0 & 1 & -1 & | & 0 \end{bmatrix}$$

$$\xrightarrow[\substack{\text{Interchange } [2], [4] \\ [1]+(-1)[2] \\ [3]+16[2] \\ [4]+(-11)[2]}]{} \begin{bmatrix} 1 & 0 & 2 & | & 1 \\ 0 & 1 & -1 & | & 0 \\ 0 & 0 & -20 & | & -6 \\ 0 & 0 & 20 & | & 6 \end{bmatrix}$$

$$\xrightarrow[\substack{-.05[3] \\ [1]+(-2)[3] \\ [2]+1[3] \\ [4]+(-20)[3]}]{} \begin{bmatrix} 1 & 0 & 0 & | & .4 \\ 0 & 1 & 0 & | & .3 \\ 0 & 0 & 1 & | & .3 \\ 0 & 0 & 0 & | & 0 \end{bmatrix}$$

$x = .4, y = .3, z = .3$

The stable distribution is $\begin{bmatrix} x \\ y \\ z \end{bmatrix} = \begin{bmatrix} .4 \\ .3 \\ .3 \end{bmatrix}$. In the long run, 40% of the people will have a strenuous workout

on a particular day.

15. The stochastic matrix is
$$\begin{array}{c} \\ D \\ R \\ H \end{array}\begin{array}{c} \begin{array}{ccc} D & R & H \end{array} \\ \left[\begin{array}{ccc} .5 & 0 & .25 \\ 0 & .5 & .25 \\ .5 & .5 & .5 \end{array}\right] \end{array}.$$

Find the stable distribution.

$$\begin{cases} x+y+z=1 \\ \left[\begin{array}{ccc} .5 & 0 & .25 \\ 0 & .5 & .25 \\ .5 & .5 & .5 \end{array}\right]\left[\begin{array}{c} x \\ y \\ z \end{array}\right]=\left[\begin{array}{c} x \\ y \\ z \end{array}\right] \end{cases}$$

$$\begin{cases} x & + & y & + & z & = & 1 \\ .5x & & & + & .25z & = & x \\ & & .5y & + & .25z & = & y \\ .5x & + & .5y & + & .5z & = & z \end{cases}$$

$$\begin{cases} x & + & y & + & z & = & 1 \\ -.5x & & & + & .25z & = & 0 \\ & & -.5y & + & .25z & = & 0 \\ .5x & + & .5y & - & .5z & = & 0 \end{cases}$$

$$\left[\begin{array}{ccc|c} 1 & 1 & 1 & 1 \\ -.5 & 0 & .25 & 0 \\ 0 & -.5 & .25 & 0 \\ .5 & .5 & -.5 & 0 \end{array}\right] \begin{array}{c} \\ 4[2] \\ \xrightarrow{4[3]} \\ 2[4] \end{array} \left[\begin{array}{ccc|c} 1 & 1 & 1 & 1 \\ -2 & 0 & 1 & 0 \\ 0 & -2 & 1 & 0 \\ 1 & 1 & -1 & 0 \end{array}\right]$$

$$\begin{array}{c} [2]+2[1] \\ \xrightarrow{} \\ [4]+(-1)[1] \end{array} \left[\begin{array}{ccc|c} 1 & 1 & 1 & 1 \\ 0 & 2 & 3 & 2 \\ 0 & -2 & 1 & 0 \\ 0 & 0 & -2 & -1 \end{array}\right]$$

$$\begin{array}{c} .5[2] \\ \xrightarrow{[1]+(-1)[2]} \\ [3]+2[2] \end{array} \left[\begin{array}{ccc|c} 1 & 0 & -.5 & 0 \\ 0 & 1 & 1.5 & 1 \\ 0 & 0 & 4 & 2 \\ 0 & 0 & -2 & -1 \end{array}\right]$$

$$\begin{array}{c} .25[3] \\ [1]+(.5)[3] \\ \xrightarrow{[2]+(-1.5)[3]} \\ [4]+2[3] \end{array} \left[\begin{array}{ccc|c} 1 & 0 & 0 & .25 \\ 0 & 1 & 0 & .25 \\ 0 & 0 & 1 & .5 \\ 0 & 0 & 0 & 0 \end{array}\right]$$

$x=.25$, $y=.25$, $z=.5$

The stable distribution is $\left[\begin{array}{c} x \\ y \\ z \end{array}\right]=\left[\begin{array}{c} .25 \\ .25 \\ .5 \end{array}\right]$.

In the long run, 25% will be dominant.

17. The stochastic matrix is $\begin{array}{cc} & \begin{array}{cc} R & S \end{array} \\ \begin{array}{c} R \\ S \end{array} & \begin{bmatrix} .1 & .6 \\ .9 & .4 \end{bmatrix} \end{array}$.

Find the stable distribution.

$$\begin{cases} x + y = 1 \\ \begin{bmatrix} .1 & .6 \\ .9 & .4 \end{bmatrix}\begin{bmatrix} x \\ y \end{bmatrix} = \begin{bmatrix} x \\ y \end{bmatrix} \end{cases}$$

$$\begin{cases} x \; + \; y = 1 \\ .1x \; + \; .6y = x \\ .9x \; + \; .4y = y \end{cases}$$

$$\begin{cases} x \; + \; y = 1 \\ -.9x \; + \; .6y = 0 \\ .9x \; - \; .6y = 0 \end{cases}$$

The second and third equations in this system are equivalent.

$$\begin{bmatrix} 1 & 1 & | & 1 \\ -.9 & .6 & | & 0 \end{bmatrix} \xrightarrow{[2]+.9[1]} \begin{bmatrix} 1 & 1 & | & 1 \\ 0 & 1.5 & | & .9 \end{bmatrix}$$

$$\xrightarrow{\frac{2}{3}[2]} \begin{bmatrix} 1 & 1 & | & 1 \\ 0 & 1 & | & .6 \end{bmatrix}$$

$$\xrightarrow{[1]+(-1)[2]} \begin{bmatrix} 1 & 0 & | & .4 \\ 0 & 1 & | & .6 \end{bmatrix}$$

$x = .4$, $y = .6$

The stable distribution is $\begin{bmatrix} x \\ y \end{bmatrix} = \begin{bmatrix} .4 \\ .6 \end{bmatrix}$.

In the log run, the daily likelihood of rain is 40% or $\frac{2}{5}$.

19. $\begin{bmatrix} .5 \\ .5 \end{bmatrix}$ is a stable distribution for the matrix

$A = \begin{bmatrix} 0 & 1 \\ 1 & 0 \end{bmatrix}$ because $.5 + .5 = 1$ and

$\begin{bmatrix} 0 & 1 \\ 1 & 0 \end{bmatrix}\begin{bmatrix} .5 \\ .5 \end{bmatrix} = \begin{bmatrix} .5 \\ .5 \end{bmatrix}$. However, given an

arbitrary initial distribution $\begin{bmatrix} \\ \end{bmatrix}_0 \neq \begin{bmatrix} .5 \\ .5 \end{bmatrix}$,

$A^n \begin{bmatrix} \\ \end{bmatrix}_0$ will not approach $\begin{bmatrix} .5 \\ .5 \end{bmatrix}$ as n gets

large, so the existence of a stable distribution

for A does not contradict the main premise of this section.

21. Calculating $[A]\verb|^|255$ gives $\begin{bmatrix} .7 & .7 \\ .3 & .3 \end{bmatrix}$.

This suggests that the stable distribution is $\begin{bmatrix} .7 \\ .3 \end{bmatrix}$.

Check: $x = .7$, $y = .3$ is indeed a solution to

the system $\begin{cases} x + y = 1 \\ \begin{bmatrix} .85 & .35 \\ .15 & .65 \end{bmatrix}\begin{bmatrix} x \\ y \end{bmatrix} = \begin{bmatrix} x \\ y \end{bmatrix} \end{cases}$.

Also $\begin{bmatrix} .85 & .35 \\ .15 & .65 \end{bmatrix}\begin{bmatrix} .7 \\ .3 \end{bmatrix} = \begin{bmatrix} .7 \\ .3 \end{bmatrix}$.

23. Calculating $[A]\verb|^|255 \blacktriangleright$ Frac gives

$\begin{bmatrix} \frac{8}{35} & \frac{8}{35} & \frac{8}{35} \\ \frac{3}{7} & \frac{3}{7} & \frac{3}{7} \\ \frac{12}{35} & \frac{12}{35} & \frac{12}{35} \end{bmatrix}$. This suggests that the stable

distribution is $\begin{bmatrix} \frac{8}{35} \\ \frac{3}{7} \\ \frac{12}{35} \end{bmatrix}$.

Check: $x = \frac{8}{35}$, $y = \frac{3}{7}$, $z = \frac{12}{35}$ is indeed a

solution to the system

$$\begin{cases} x + y + z = 1 \\ \begin{bmatrix} .1 & .4 & .1 \\ .3 & .2 & .8 \\ .6 & .4 & .1 \end{bmatrix}\begin{bmatrix} x \\ y \\ z \end{bmatrix} = \begin{bmatrix} x \\ y \\ z \end{bmatrix} \end{cases}.$$

Also $\begin{bmatrix} .1 & .4 & .1 \\ .3 & .2 & .8 \\ .6 & .4 & .1 \end{bmatrix}\begin{bmatrix} \frac{8}{35} \\ \frac{3}{7} \\ \frac{12}{35} \end{bmatrix} = \begin{bmatrix} \frac{8}{35} \\ \frac{3}{7} \\ \frac{12}{35} \end{bmatrix}$.

Exercises 3

1. No; states 1 and 2 are absorbing states, but states 3 and 4 do not lead to absorbing states.

3. Yes; state 1 is an absorbing state, and state 3 leads to state 1. Furthermore, it is possible to go from state 2 to state 1 through an intermediate step (state 2 to state 3 to state 1).

5.
$$\left[\begin{array}{cc|c} 1 & 0 & .3 \\ 0 & 1 & .2 \\ \hline 0 & 0 & .5 \end{array}\right] = \left[\begin{array}{c|c} I & S \\ \hline 0 & R \end{array}\right]$$

$$R = [.5]; \; S = \begin{bmatrix} .3 \\ .2 \end{bmatrix};$$

$$F = (I - R)^{-1} = [1 - .5]^{-1} = [2]$$

Find the stable matrix:

$$S(I - R)^{-1} = \begin{bmatrix} .3 \\ .2 \end{bmatrix}[2] = \begin{bmatrix} .6 \\ .4 \end{bmatrix}$$

$$\left[\begin{array}{c|c} I & S(I - R)^{-1} \\ \hline 0 & 0 \end{array}\right] = \left[\begin{array}{cc|c} 1 & 0 & .6 \\ 0 & 1 & .4 \\ \hline 0 & 0 & 0 \end{array}\right]$$

7.
$$\left[\begin{array}{cc|cc} 1 & 0 & .1 & 0 \\ 0 & 1 & .5 & .2 \\ \hline 0 & 0 & .3 & .6 \\ 0 & 0 & .1 & .2 \end{array}\right] = \left[\begin{array}{c|c} I & S \\ \hline 0 & R \end{array}\right]$$

$$R = \begin{bmatrix} .3 & .6 \\ .1 & .2 \end{bmatrix}; \; S = \begin{bmatrix} .1 & 0 \\ .5 & .2 \end{bmatrix}$$

$$I - R = \begin{bmatrix} 1 & 0 \\ 0 & 1 \end{bmatrix} - \begin{bmatrix} .3 & .6 \\ .1 & .2 \end{bmatrix} = \begin{bmatrix} .7 & -.6 \\ -.1 & .8 \end{bmatrix}$$

$$= \begin{bmatrix} a & b \\ c & d \end{bmatrix}$$

$$\Delta = ad - bc = (.7)(.8) - (-.6)(-.1) = .5$$

$$F = (I - R)^{-1} = \begin{bmatrix} \frac{d}{\Delta} & -\frac{b}{\Delta} \\ -\frac{c}{\Delta} & \frac{a}{\Delta} \end{bmatrix} = \begin{bmatrix} \frac{.8}{.5} & \frac{-.6}{.5} \\ -\frac{.1}{.5} & \frac{.7}{.5} \end{bmatrix}$$

$$= \begin{bmatrix} 1.6 & 1.2 \\ .2 & 1.4 \end{bmatrix}$$

Find the stable matrix:

$$S(I - R)^{-1} = \begin{bmatrix} .1 & 0 \\ .5 & .2 \end{bmatrix}\begin{bmatrix} 1.6 & 1.2 \\ .2 & 1.4 \end{bmatrix}$$

$$= \begin{bmatrix} .16 & .12 \\ .84 & .88 \end{bmatrix}$$

$$\left[\begin{array}{c|c} I & S(I - R)^{-1} \\ \hline 0 & 0 \end{array}\right] = \left[\begin{array}{cc|cc} 1 & 0 & .16 & .12 \\ 0 & 1 & .84 & .88 \\ \hline 0 & 0 & 0 & 0 \\ 0 & 0 & 0 & 0 \end{array}\right]$$

9.
$$\left[\begin{array}{ccc|cc} 1 & 0 & 0 & .1 & .2 \\ 0 & 1 & 0 & .3 & 0 \\ 0 & 0 & 1 & 0 & .2 \\ \hline 0 & 0 & 0 & .5 & 0 \\ 0 & 0 & 0 & .1 & .6 \end{array}\right] = \left[\begin{array}{c|c} I & S \\ \hline 0 & R \end{array}\right]$$

$$R = \begin{bmatrix} .5 & 0 \\ .1 & .6 \end{bmatrix}; \; S = \begin{bmatrix} .1 & .2 \\ .3 & 0 \\ 0 & .2 \end{bmatrix}$$

Find the fundamental matrix:

$$I - R = \begin{bmatrix} 1 & 0 \\ 0 & 1 \end{bmatrix} - \begin{bmatrix} .5 & 0 \\ .1 & .6 \end{bmatrix} = \begin{bmatrix} .5 & 0 \\ -.1 & .4 \end{bmatrix}$$

$$= \begin{bmatrix} a & b \\ c & d \end{bmatrix}$$

$$\Delta = ad - bc = (.5)(.4) - (0)(-.1) = .2$$

$$F = (I - R)^{-1} = \begin{bmatrix} \frac{d}{\Delta} & -\frac{b}{\Delta} \\ -\frac{c}{\Delta} & \frac{a}{\Delta} \end{bmatrix} = \begin{bmatrix} \frac{.4}{.2} & -\frac{0}{.2} \\ -\frac{.1}{.2} & \frac{.5}{.2} \end{bmatrix}$$

$$= \begin{bmatrix} 2 & 0 \\ .5 & 2.5 \end{bmatrix}$$

Find the stable matrix:

$$S(I - R)^{-1} = \begin{bmatrix} .1 & .2 \\ .3 & 0 \\ 0 & .2 \end{bmatrix}\begin{bmatrix} 2 & 0 \\ .5 & 2.5 \end{bmatrix}$$

$$= \begin{bmatrix} .3 & .5 \\ .6 & 0 \\ .1 & .5 \end{bmatrix}$$

$$\left[\begin{array}{c|c} I & S(I - R)^{-1} \\ \hline 0 & 0 \end{array}\right] = \left[\begin{array}{ccc|cc} 1 & 0 & 0 & .3 & .5 \\ 0 & 1 & 0 & .6 & 0 \\ 0 & 0 & 1 & .1 & .5 \\ \hline 0 & 0 & 0 & 0 & 0 \\ 0 & 0 & 0 & 0 & 0 \end{array}\right]$$

11. If the gambler begins with $2, he should have $1 for an expected number of .79 plays.

13. a.

$$\begin{array}{c} \\ D \\ G \\ F \\ S \end{array} \begin{array}{cccc} D & G & F & S \\ \left[\begin{array}{cc|cc} 1 & 0 & .2 & .1 \\ 0 & 1 & 0 & .9 \\ \hline 0 & 0 & 0 & 0 \\ 0 & 0 & .8 & 0 \end{array}\right] \end{array}$$

b. $R = \begin{bmatrix} 0 & 0 \\ .8 & 0 \end{bmatrix}$; $S = \begin{bmatrix} .2 & .1 \\ 0 & .9 \end{bmatrix}$

$$I - R = \begin{bmatrix} 1 & 0 \\ 0 & 1 \end{bmatrix} - \begin{bmatrix} 0 & 0 \\ .8 & 0 \end{bmatrix} = \begin{bmatrix} 1 & 0 \\ -.8 & 1 \end{bmatrix}$$

$$= \begin{bmatrix} a & b \\ c & d \end{bmatrix}$$

$\Delta = ad - bc = (1)(1) - (0)(-.8) = 1$

$$(I - R)^{-1} = \begin{bmatrix} \frac{d}{\Delta} & -\frac{b}{\Delta} \\ -\frac{c}{\Delta} & \frac{a}{\Delta} \end{bmatrix} = \begin{bmatrix} \frac{1}{1} & -\frac{0}{1} \\ -\frac{.8}{1} & \frac{1}{1} \end{bmatrix}$$

$$= \begin{bmatrix} 1 & 0 \\ .8 & 1 \end{bmatrix}$$

$$S(I - R)^{-1} = \begin{bmatrix} .2 & .1 \\ 0 & .9 \end{bmatrix}\begin{bmatrix} 1 & 0 \\ .8 & 1 \end{bmatrix}$$

$$= \begin{bmatrix} .28 & .1 \\ .72 & .9 \end{bmatrix}$$

$$\begin{bmatrix} I & S(I-R)^{-1} \\ \hline 0 & 0 \end{bmatrix} = \begin{array}{c} D \\ G \\ F \\ S \end{array} \begin{array}{cccc} D & G & F & S \\ \left[\begin{array}{cc|cc} 1 & 0 & .28 & .1 \\ 0 & 1 & .72 & .9 \\ \hline 0 & 0 & 0 & 0 \\ 0 & 0 & 0 & 0 \end{array}\right] \end{array}$$

c. From the *F* column of the stable matrix, the probability that a freshman will eventually graduate is .72.

d. The fundamental matrix is

$$F = (I - R)^{-1} = \begin{array}{c} \\ F \\ S \end{array} \begin{array}{cc} F & S \\ \begin{bmatrix} 1 & 0 \\ .8 & 1 \end{bmatrix} \end{array}.$$

Add the numbers in the *F* column: $1 + .8 = 1.8$.
The student will attend an expected number of 1.8 years.

15. First, arrange the matrix so that the absorbing states come first and calculate the fundamental and stable matrices.

$$\begin{array}{c} \\ \text{Paid} \\ \text{Bad} \\ \leq 30 \\ < 60 \end{array} \begin{array}{cccc} \text{Paid} & \text{Bad} & \leq 30 & < 60 \\ \left[\begin{array}{cc|cc} 1 & 0 & .4 & .1 \\ 0 & 1 & 0 & .1 \\ \hline 0 & 0 & .4 & .4 \\ 0 & 0 & .2 & .4 \end{array}\right] \end{array} = \begin{bmatrix} I & S \\ \hline 0 & R \end{bmatrix}$$

$$R = \begin{bmatrix} .4 & .4 \\ .2 & .4 \end{bmatrix}; S = \begin{bmatrix} .4 & .1 \\ 0 & .1 \end{bmatrix}$$

$$I - R = \begin{bmatrix} 1 & 0 \\ 0 & 1 \end{bmatrix} - \begin{bmatrix} .4 & .4 \\ .2 & .4 \end{bmatrix} = \begin{bmatrix} .6 & -.4 \\ -.2 & .6 \end{bmatrix}$$

$$= \begin{bmatrix} a & b \\ c & d \end{bmatrix}$$

$\Delta = ad - bc = (.6)(.6) - (-.4)(-.2) = .28$

$$F = (I - R)^{-1} = \begin{bmatrix} \frac{d}{\Delta} & -\frac{b}{\Delta} \\ -\frac{c}{\Delta} & \frac{a}{\Delta} \end{bmatrix}$$

$$= \begin{bmatrix} \frac{.6}{.28} & -\frac{-.4}{.28} \\ -\frac{-.2}{.28} & \frac{.6}{.28} \end{bmatrix} = \begin{bmatrix} \frac{15}{7} & \frac{10}{7} \\ \frac{5}{7} & \frac{15}{7} \end{bmatrix}$$

$$S(I - R)^{-1} = \begin{bmatrix} .4 & .1 \\ 0 & .1 \end{bmatrix}\begin{bmatrix} \frac{15}{7} & \frac{10}{7} \\ \frac{5}{7} & \frac{15}{7} \end{bmatrix}$$

$$= \begin{bmatrix} \frac{13}{14} & \frac{11}{14} \\ \frac{1}{14} & \frac{3}{14} \end{bmatrix}$$

The stable matrix is

$$\begin{array}{c} \\ \text{Paid} \\ \text{Bad} \\ \leq 30 \\ < 60 \end{array} \begin{array}{cccc} \text{Paid} & \text{Bad} & \leq 30 & < 60 \\ \left[\begin{array}{cc|cc} 1 & 0 & \frac{13}{14} & \frac{11}{14} \\ 0 & 1 & \frac{1}{14} & \frac{3}{14} \\ \hline 0 & 0 & 0 & 0 \\ 0 & 0 & 0 & 0 \end{array}\right] \end{array}.$$

a. From the stable matrix, the probability of an account eventually being paid off is $\frac{13}{14}$ if it is currently at most 30 days overdue and $\frac{11}{14}$ if it is less than 60 days overdue (but more than 30 days overdue).

b. The fundamental matrix is

$$
\begin{array}{cc}
 & \leq 30 \quad < 60 \\
\begin{array}{c} \leq 30 \\ < 60 \end{array} &
\begin{bmatrix} \frac{15}{7} & \frac{10}{7} \\ \frac{5}{7} & \frac{15}{7} \end{bmatrix}
\end{array}. \text{ Add the numbers}
$$

in the first column: $\dfrac{15}{7} + \dfrac{5}{7} = \dfrac{20}{7}$. An account that is overdue at most 30 days is expected to reach an absorbing state (paid or bad) after $\dfrac{20}{7}$ months.

c. From the stable matrix, about $\dfrac{13}{14}$ of the "≤30 day" debt will be paid, and about $\dfrac{11}{14}$ of the "<60 day" debt will be paid.

$\dfrac{13}{14}(\$2000) + \dfrac{11}{14}(\$5000) \approx \$5786$.

About \$5786 will be paid and about \$1214 will become bad debt.

17. First, find the absorbing stochastic matrix and calculate the fundamental and stable matrices.

$$
\begin{array}{c}
 \quad\;\; \$0 \;\; \$4 \;\; \$1 \;\; \$2 \;\; \$3 \\
\begin{array}{c} \$0 \\ \$4 \\ \$1 \\ \$2 \\ \$3 \end{array}
\left[\begin{array}{cc|ccc}
1 & 0 & \frac{1}{2} & 0 & 0 \\
0 & 1 & 0 & 0 & \frac{1}{2} \\
\hline
0 & 0 & 0 & \frac{1}{2} & 0 \\
0 & 0 & \frac{1}{2} & 0 & \frac{1}{2} \\
0 & 0 & 0 & \frac{1}{2} & 0
\end{array}\right] =
\begin{bmatrix} I & S \\ 0 & R \end{bmatrix}
\end{array}
$$

$$
R = \begin{bmatrix} 0 & \frac{1}{2} & 0 \\ \frac{1}{2} & 0 & \frac{1}{2} \\ 0 & \frac{1}{2} & 0 \end{bmatrix};\;\;
S = \begin{bmatrix} \frac{1}{2} & 0 & 0 \\ 0 & 0 & \frac{1}{2} \end{bmatrix}
$$

$$
I - R = \begin{bmatrix} 1 & 0 & 0 \\ 0 & 1 & 0 \\ 0 & 0 & 1 \end{bmatrix} - \begin{bmatrix} 0 & \frac{1}{2} & 0 \\ \frac{1}{2} & 0 & \frac{1}{2} \\ 0 & \frac{1}{2} & 0 \end{bmatrix}
$$

$$
= \begin{bmatrix} 1 & -\frac{1}{2} & 0 \\ -\frac{1}{2} & 1 & -\frac{1}{2} \\ 0 & -\frac{1}{2} & 1 \end{bmatrix}
$$

$$
F = (I - R)^{-1} = \begin{bmatrix} 1 & -\frac{1}{2} & 0 \\ -\frac{1}{2} & 1 & -\frac{1}{2} \\ 0 & -\frac{1}{2} & 1 \end{bmatrix}^{-1}
$$

$$
= \begin{bmatrix} \frac{3}{2} & 1 & \frac{1}{2} \\ 1 & 2 & 1 \\ \frac{1}{2} & 1 & \frac{3}{2} \end{bmatrix}
$$

$$
S(I - R)^{-1} = \begin{bmatrix} \frac{1}{2} & 0 & 0 \\ 0 & 0 & \frac{1}{2} \end{bmatrix}\begin{bmatrix} \frac{3}{2} & 1 & \frac{1}{2} \\ 1 & 2 & 1 \\ \frac{1}{2} & 1 & \frac{3}{2} \end{bmatrix}
$$

$$
= \begin{bmatrix} \frac{3}{4} & \frac{1}{2} & \frac{1}{4} \\ \frac{1}{4} & \frac{1}{2} & \frac{3}{4} \end{bmatrix}
$$

The stable matrix is

$$
\begin{array}{c}
 \quad\;\; \$0 \;\; \$4 \;\; \$1 \;\; \$2 \;\; \$3 \\
\begin{array}{c} \$0 \\ \$4 \\ \$1 \\ \$2 \\ \$3 \end{array}
\left[\begin{array}{cc|ccc}
0 & 1 & \frac{3}{4} & \frac{1}{2} & \frac{1}{4} \\
1 & 0 & \frac{1}{4} & \frac{1}{2} & \frac{3}{4} \\
\hline
0 & 0 & 0 & 0 & 0 \\
0 & 0 & 0 & 0 & 0 \\
0 & 0 & 0 & 0 & 0
\end{array}\right].
\end{array}
$$

a. From the top row of the stable matrix, the probability of eventually going broke is $\dfrac{3}{4}$ if he starts with \$1, $\dfrac{1}{2}$ if he starts with \$2, and $\dfrac{1}{4}$ if he starts with \$3.

b. The fundamental matrix is

$$
\begin{array}{c}
 \quad\; \$1 \;\; \$2 \;\; \$3 \\
\begin{array}{c} \$1 \\ \$2 \\ \$3 \end{array}
\begin{bmatrix} \frac{3}{2} & 1 & \frac{1}{2} \\ 1 & 2 & 1 \\ \frac{1}{2} & 1 & \frac{3}{2} \end{bmatrix}.
\end{array}
$$

Add the entries in the middle column: $1 + 2 + 1 = 4$

Starting with \$2, he will play for an expected number of 4 times.

19. Calculating [A] ^ 255 ▶ Frac gives

$$\begin{bmatrix} 1 & 0 & \frac{666}{917} & \frac{75}{131} & \frac{250}{917} \\ 0 & 1 & \frac{251}{917} & \frac{56}{131} & \frac{667}{917} \\ 0 & 0 & 0 & 0 & 0 \\ 0 & 0 & 0 & 0 & 0 \\ 0 & 0 & 0 & 0 & 0 \end{bmatrix}.$$

Let $[B] = S = \begin{bmatrix} .5 & .4 & .2 \\ 0 & .3 & .7 \end{bmatrix}$ and

$[C] = R = \begin{bmatrix} 0 & .2 & .1 \\ .3 & 0 & 0 \\ .2 & .1 & 0 \end{bmatrix}$. Then calculating

$[B] * (\text{identity }(3) - [C])^{-1}$ ▶ Frac gives

$\begin{bmatrix} \frac{666}{917} & \frac{75}{131} & \frac{250}{917} \\ \frac{251}{917} & \frac{56}{131} & \frac{667}{917} \end{bmatrix}$. Therefore, the matrix

given above is the exact stable matrix.

Chapter 8 Supplementary Exercises

1. Stochastic, neither; the matrix is stochastic because it is square, the entries are all ≥ 0, and the sum of the entries in each column is 1. It is not regular because all powers of the matrix include zero entries. The first state is an absorbing state, but the matrix is not an absorbing matrix because an object that begins in the third or fourth state will always remain within these two (nonabsorbing) states.

2. Stochastic, regular; the matrix is stochastic because it is square, the entries are all ≥ 0, and the sum of the entries in each column is 1. It is regular because it has no zero entries.

3. Stochastic, regular; the matrix is stochastic because it is square, the entries are all ≥ 0, and the sum of the entries in each column is 1. It is regular because the second power, $\begin{bmatrix} .3 & .21 \\ .7 & .79 \end{bmatrix}$, contains no zero entries.

4. Stochastic, absorbing; the matrix is stochastic because it is square, the entries are all ≥ 0, and the sum of the entries in each column is 1. It is absorbing because the first and second states are absorbing states, and it is possible for an object to get from the third state to an absorbing state.

5. Not stochastic because the sum of the entries in the middle column is not 1.

6. Stochastic, absorbing; the matrix is stochastic because it is square, the entries are all ≥ 0, and the sum of the entries in each column is 1. It is absorbing because the first and second states are absorbing states and it is possible to get from the third state (indirectly) or the fourth state (directly) to an absorbing state.

7. $\begin{cases} x + y = 1 \\ \begin{bmatrix} .6 & .5 \\ .4 & .5 \end{bmatrix} \begin{bmatrix} x \\ y \end{bmatrix} = \begin{bmatrix} x \\ y \end{bmatrix} \end{cases}$

$\begin{cases} x + y = 1 \\ .6x + .5y = x \\ .4x + .5y = y \end{cases}$

$\begin{cases} x + y = 1 \\ -.4x + .5y = 0 \\ .4x - .5y = 0 \end{cases}$

The second and third equations in this system are equivalent.

$\begin{bmatrix} 1 & 1 & | & 1 \\ -.4 & .5 & | & 0 \end{bmatrix} \xrightarrow{[2]+.4[1]} \begin{bmatrix} 1 & 1 & | & 1 \\ 0 & .9 & | & .4 \end{bmatrix}$

$\xrightarrow{\frac{10}{9}[2]} \begin{bmatrix} 1 & 1 & | & 1 \\ 0 & 1 & | & \frac{4}{9} \end{bmatrix}$

$\xrightarrow{[1]+(-1)[2]} \begin{bmatrix} 1 & 0 & | & \frac{5}{9} \\ 0 & 1 & | & \frac{4}{9} \end{bmatrix}$

$x = \dfrac{5}{9},\ y = \dfrac{4}{9}$

The stable distribution is $\begin{bmatrix} \frac{5}{9} \\ \frac{4}{9} \end{bmatrix}$.

8. $\begin{bmatrix} 1 & 0 & 0 & \frac{1}{8} & \frac{1}{4} \\ 0 & 1 & 0 & \frac{1}{8} & 0 \\ 0 & 0 & 1 & 0 & \frac{1}{4} \\ \hline 0 & 0 & 0 & \frac{1}{4} & \frac{1}{2} \\ 0 & 0 & 0 & \frac{1}{2} & 0 \end{bmatrix} = \begin{bmatrix} I & S \\ \hline 0 & R \end{bmatrix}$

$R = \begin{bmatrix} \frac{1}{4} & \frac{1}{2} \\ \frac{1}{2} & 0 \end{bmatrix}; \quad S = \begin{bmatrix} \frac{1}{8} & \frac{1}{4} \\ \frac{1}{8} & 0 \\ 0 & \frac{1}{4} \end{bmatrix}$

$I - R = \begin{bmatrix} 1 & 0 \\ 0 & 1 \end{bmatrix} - \begin{bmatrix} \frac{1}{4} & \frac{1}{2} \\ \frac{1}{2} & 0 \end{bmatrix} = \begin{bmatrix} \frac{3}{4} & -\frac{1}{2} \\ -\frac{1}{2} & 1 \end{bmatrix} = \begin{bmatrix} a & b \\ c & d \end{bmatrix}$

$\Delta = ad - bc = \left(\frac{3}{4}\right)(1) - \left(-\frac{1}{2}\right)\left(-\frac{1}{2}\right) = \frac{1}{2}$

$(I - R)^{-1} = \begin{bmatrix} \frac{d}{\Delta} & -\frac{b}{\Delta} \\ -\frac{c}{\Delta} & \frac{a}{\Delta} \end{bmatrix} = \begin{bmatrix} \frac{1}{1/2} & -\frac{-1/2}{1/2} \\ -\frac{-1/2}{1/2} & \frac{3/4}{1/2} \end{bmatrix} = \begin{bmatrix} 2 & 1 \\ 1 & \frac{3}{2} \end{bmatrix}$

$S(I - R)^{-1} = \begin{bmatrix} \frac{1}{8} & \frac{1}{4} \\ \frac{1}{8} & 0 \\ 0 & \frac{1}{4} \end{bmatrix} \begin{bmatrix} 2 & 1 \\ 1 & \frac{3}{2} \end{bmatrix} = \begin{bmatrix} \frac{1}{2} & \frac{1}{2} \\ \frac{1}{4} & \frac{1}{8} \\ \frac{1}{4} & \frac{3}{8} \end{bmatrix}$

$\begin{bmatrix} I & S(I-R)^{-1} \\ \hline 0 & 0 \end{bmatrix} = \begin{bmatrix} 1 & 0 & 0 & \frac{1}{2} & \frac{1}{2} \\ 0 & 1 & 0 & \frac{1}{4} & \frac{1}{8} \\ 0 & 0 & 1 & \frac{1}{4} & \frac{3}{8} \\ \hline 0 & 0 & 0 & 0 & 0 \\ 0 & 0 & 0 & 0 & 0 \end{bmatrix}$

9. a.

$\begin{array}{c} \\ H \\ M \\ L \end{array} \begin{array}{ccc} H & M & L \end{array} \\ \begin{bmatrix} .5 & .4 & .3 \\ .4 & .3 & .5 \\ .1 & .3 & .2 \end{bmatrix}$

b. $\begin{bmatrix} .5 & .4 & .3 \\ .4 & .3 & .5 \\ .1 & .3 & .2 \end{bmatrix} \begin{bmatrix} .1 \\ .6 \\ .3 \end{bmatrix} = \begin{bmatrix} .38 \\ .37 \\ .25 \end{bmatrix}$

38% of the children of the current generation will have high incomes.

c. Find the stable distribution.

$$\begin{cases} x + y + z = 1 \\ \begin{bmatrix} .5 & .4 & .3 \\ .4 & .3 & .5 \\ .1 & .3 & .2 \end{bmatrix} \begin{bmatrix} x \\ y \\ z \end{bmatrix} = \begin{bmatrix} x \\ y \\ z \end{bmatrix} \end{cases}$$

$$\begin{cases} x & + & y & + & z & = & 1 \\ .5x & + & .4y & + & .3z & = & x \\ .4x & + & .3y & + & .5z & = & y \\ .1x & + & .3y & + & .2z & = & z \end{cases}$$

$$\begin{cases} x & + & y & + & z & = & 1 \\ -.5x & + & .4y & + & .3z & = & 0 \\ .4x & - & .7y & + & .5z & = & 0 \\ .1x & + & .3y & - & .8z & = & 0 \end{cases}$$

$$\begin{bmatrix} 1 & 1 & 1 & | & 1 \\ -.5 & .4 & .3 & | & 0 \\ .4 & -.7 & .5 & | & 0 \\ .1 & .3 & -.8 & | & 0 \end{bmatrix} \begin{matrix} \\ 10[2] \\ 10[3] \\ 10[4] \end{matrix} \longrightarrow \begin{bmatrix} 1 & 1 & 1 & | & 1 \\ -5 & 4 & 3 & | & 0 \\ 4 & -7 & 5 & | & 0 \\ 1 & 3 & -8 & | & 0 \end{bmatrix}$$

$$\begin{matrix} [2] + 5[1] \\ [3] + (-4)[1] \\ [4] + (-1)[1] \end{matrix} \longrightarrow \begin{bmatrix} 1 & 1 & 1 & | & 1 \\ 0 & 9 & 8 & | & 5 \\ 0 & -11 & 1 & | & -4 \\ 0 & 2 & -9 & | & -1 \end{bmatrix}$$

$$\begin{matrix} \frac{1}{9}[2] \\ [1] + (-1)[2] \\ [3] + (11)[2] \\ [4] + (-2)[2] \end{matrix} \longrightarrow \begin{bmatrix} 1 & 0 & \frac{1}{9} & | & \frac{4}{9} \\ 0 & 1 & \frac{8}{9} & | & \frac{5}{9} \\ 0 & 0 & \frac{97}{9} & | & \frac{19}{9} \\ 0 & 0 & -\frac{97}{9} & | & -\frac{19}{9} \end{bmatrix}$$

$$\begin{matrix} \frac{9}{97}[3] \\ [1] + \left(-\frac{1}{9}\right)[3] \\ [2] + \left(-\frac{8}{9}\right)[3] \\ [4] + \frac{97}{9}[3] \end{matrix} \longrightarrow \begin{bmatrix} 1 & 0 & 0 & | & \frac{41}{97} \\ 0 & 1 & 0 & | & \frac{37}{97} \\ 0 & 0 & 1 & | & \frac{19}{97} \\ 0 & 0 & 0 & | & 0 \end{bmatrix}$$

$$x = \frac{41}{97}, \ y = \frac{37}{97}, \ z = \frac{19}{97}.$$

The stable distribution is $\begin{bmatrix} \frac{41}{97} \\ \frac{37}{97} \\ \frac{19}{97} \end{bmatrix}$. In the long run, $\frac{19}{97}$ of the population will have low incomes.

10. a.

$$\begin{array}{cc} & \begin{array}{cc} P & N \end{array} \\ \begin{array}{c} P \\ N \end{array} & \begin{bmatrix} .8 & .3 \\ .2 & .7 \end{bmatrix} \end{array}$$

b. $A^2 \begin{bmatrix} \\ \\ \end{bmatrix}_0 = \begin{bmatrix} .8 & .3 \\ .2 & .7 \end{bmatrix}\begin{bmatrix} .8 & .3 \\ .2 & .7 \end{bmatrix}\begin{bmatrix} 1 \\ 0 \end{bmatrix}_0 = \begin{bmatrix} .7 \\ .3 \end{bmatrix}_2$

30% will need adjusting after 2 days.

c. Find the stable distribution.

$$\begin{cases} x + y = 1 \\ \begin{bmatrix} .8 & .3 \\ .2 & .7 \end{bmatrix}\begin{bmatrix} x \\ y \end{bmatrix} = \begin{bmatrix} x \\ y \end{bmatrix} \end{cases}$$

$$\begin{cases} x & + & y & = & 1 \\ .8x & + & .3y & = & x \\ .2x & + & .7y & = & y \end{cases}$$

$$\begin{cases} x & + & y & = & 1 \\ -.2x & + & .3y & = & 0 \\ .2x & - & .3y & = & 0 \end{cases}$$

The second and third equations in this system are equivalent.

$$\begin{bmatrix} 1 & 1 & | & 1 \\ -.2 & .3 & | & 0 \end{bmatrix} \xrightarrow{[2]+.2[1]} \begin{bmatrix} 1 & 1 & | & 1 \\ 0 & .5 & | & .2 \end{bmatrix}$$

$$\xrightarrow{2[2]} \begin{bmatrix} 1 & 1 & | & 1 \\ 0 & 1 & | & .4 \end{bmatrix}$$

$$\xrightarrow{[1]+(-1)[2]} \begin{bmatrix} 1 & 0 & | & .6 \\ 0 & 1 & | & .4 \end{bmatrix}$$

$x = .6, y = .4$

The stable distribution is $\begin{bmatrix} .6 \\ .4 \end{bmatrix}$. In the long run, 60% will be properly adjusted.

11. $\begin{bmatrix} 1 & 0 & \frac{1}{6} & \frac{1}{2} & \frac{2}{5} \\ 0 & 1 & 0 & 0 & \frac{2}{5} \\ 0 & 0 & 0 & 0 & 0 \\ 0 & 0 & \frac{2}{3} & \frac{1}{2} & 0 \\ 0 & 0 & \frac{1}{6} & 0 & \frac{1}{5} \end{bmatrix} = \begin{bmatrix} I & | & S \\ \hline 0 & | & R \end{bmatrix}$

$R = \begin{bmatrix} 0 & 0 & 0 \\ \frac{2}{3} & \frac{1}{2} & 0 \\ \frac{1}{6} & 0 & \frac{1}{5} \end{bmatrix}; \ S = \begin{bmatrix} \frac{1}{6} & \frac{1}{2} & \frac{2}{5} \\ 0 & 0 & \frac{2}{5} \end{bmatrix}$

$$I - R = \begin{bmatrix} 1 & 0 & 0 \\ 0 & 1 & 0 \\ 0 & 0 & 1 \end{bmatrix} - \begin{bmatrix} 0 & 0 & 0 \\ \frac{2}{3} & \frac{1}{2} & 0 \\ \frac{1}{6} & 0 & \frac{1}{5} \end{bmatrix} = \begin{bmatrix} 1 & 0 & 0 \\ -\frac{2}{3} & \frac{1}{2} & 0 \\ -\frac{1}{6} & 0 & \frac{4}{5} \end{bmatrix}$$

Use the Gauss-Jordan method to find $(I - R)^{-1}$.

$$\begin{bmatrix} 1 & 0 & 0 & | & 1 & 0 & 0 \\ -\frac{2}{3} & \frac{1}{2} & 0 & | & 0 & 1 & 0 \\ -\frac{1}{6} & 0 & \frac{4}{5} & | & 0 & 0 & 1 \end{bmatrix} \xrightarrow[{[3]+\frac{1}{6}[1]}]{[2]+\frac{2}{3}[1]} \begin{bmatrix} 1 & 0 & 0 & | & 1 & 0 & 0 \\ 0 & \frac{1}{2} & 0 & | & \frac{2}{3} & 1 & 0 \\ 0 & 0 & \frac{4}{5} & | & \frac{1}{6} & 0 & 1 \end{bmatrix}$$

$$\xrightarrow[{\frac{5}{4}[3]}]{2[2]} \begin{bmatrix} 1 & 0 & 0 & | & 1 & 0 & 0 \\ 0 & 1 & 0 & | & \frac{4}{3} & 2 & 0 \\ 0 & 0 & 1 & | & \frac{5}{24} & 0 & \frac{5}{4} \end{bmatrix}$$

$$S(I - R)^{-1} = \begin{bmatrix} \frac{1}{6} & \frac{1}{2} & \frac{2}{5} \\ 0 & 0 & \frac{2}{5} \end{bmatrix} \begin{bmatrix} 1 & 0 & 0 \\ \frac{4}{3} & 2 & 0 \\ \frac{5}{24} & 0 & \frac{5}{4} \end{bmatrix} = \begin{bmatrix} \frac{11}{12} & 1 & \frac{1}{2} \\ \frac{1}{12} & 0 & \frac{1}{2} \end{bmatrix}$$

$$\begin{bmatrix} I & | & S(I - R)^{-1} \\ \hline 0 & | & 0 \end{bmatrix} = \begin{bmatrix} 1 & 0 & | & \frac{11}{12} & 1 & \frac{1}{2} \\ 0 & 1 & | & \frac{1}{12} & 0 & \frac{1}{2} \\ \hline 0 & 0 & | & 0 & 0 & 0 \\ 0 & 0 & | & 0 & 0 & 0 \\ 0 & 0 & | & 0 & 0 & 0 \end{bmatrix}$$

12. a.

	I	II	III	IV
I	1	0	0	$\frac{1}{4}$
II	0	1	$\frac{1}{3}$	$\frac{1}{4}$
III	0	0	0	$\frac{1}{2}$
IV	0	0	$\frac{2}{3}$	0

b. The only way the mouse can find the cheese in exactly two minutes is to go to room III after the first minute and then to room II the next minute. The probability of this occurring is $\frac{1}{2} \cdot \frac{1}{3} = \frac{1}{6}$. The probability that the mouse will find the cheese in exactly 1 minute is $\frac{1}{4}$. Thus the probability of the mouse finding the cheese after two minutes is $\frac{1}{6} + \frac{1}{4} = \frac{5}{12}$.

c. Find the stable matrix.

$$R = \begin{bmatrix} 0 & \frac{1}{2} \\ \frac{2}{3} & 0 \end{bmatrix}, \; S = \begin{bmatrix} 0 & \frac{1}{4} \\ \frac{1}{3} & \frac{1}{4} \end{bmatrix}$$

$$I - R = \begin{bmatrix} 1 & 0 \\ 0 & 1 \end{bmatrix} - \begin{bmatrix} 0 & \frac{1}{2} \\ \frac{2}{3} & 0 \end{bmatrix} = \begin{bmatrix} 1 & -\frac{1}{2} \\ -\frac{2}{3} & 1 \end{bmatrix}$$

$$= \begin{bmatrix} a & b \\ c & d \end{bmatrix}$$

$$\Delta = ad - bc = (1)(1) - \left(-\frac{1}{2}\right)\left(-\frac{2}{3}\right) = \frac{2}{3}$$

$$F = (I - R)^{-1} = \begin{bmatrix} \frac{d}{\Delta} & -\frac{b}{\Delta} \\ -\frac{c}{\Delta} & \frac{a}{\Delta} \end{bmatrix}$$

$$= \begin{bmatrix} \frac{1}{2/3} & -\frac{-1/2}{2/3} \\ -\frac{-2/3}{2/3} & \frac{1}{2/3} \end{bmatrix} = \begin{bmatrix} \frac{3}{2} & \frac{3}{4} \\ 1 & \frac{3}{2} \end{bmatrix}$$

$$S(I - R)^{-1} = \begin{bmatrix} 0 & \frac{1}{4} \\ \frac{1}{3} & \frac{1}{4} \end{bmatrix} \begin{bmatrix} \frac{3}{2} & \frac{3}{4} \\ 1 & \frac{3}{2} \end{bmatrix}$$

$$= \begin{bmatrix} \frac{1}{4} & \frac{3}{8} \\ \frac{3}{4} & \frac{5}{8} \end{bmatrix}$$

The stable matrix is $\left[\begin{array}{c|c} I & S(I-R)^{-1} \\ \hline 0 & 0 \end{array} \right] = \begin{array}{c} \\ I \\ II \\ III \\ IV \end{array} \begin{array}{cccc} I & II & III & IV \\ \left[\begin{array}{cc|cc} 1 & 0 & \frac{1}{4} & \frac{3}{8} \\ 0 & 1 & \frac{3}{4} & \frac{5}{8} \\ \hline 0 & 0 & 0 & 0 \\ 0 & 0 & 0 & 0 \end{array} \right] \end{array}$

From the fourth column, if he starts in room IV the probability of finding cheese in the long run is $\frac{5}{8}$.

d. The fundamental matrix is $\begin{array}{c} \\ III \\ IV \end{array} \begin{array}{c} III IV \\ \begin{bmatrix} \frac{3}{2} & \frac{3}{4} \\ 1 & \frac{3}{2} \end{bmatrix} \end{array}$ Add the entries in the column for room III: $\frac{3}{2} + 1 = \frac{5}{2}$.

A mouse that starts in room III will spend an expected number of $\frac{5}{2}$ minutes before finding the cheese or being trapped.

13. (c) is the correct choice because it satisfies the system $\begin{cases} x + y + z = 1 \\ \begin{bmatrix} .4 & .4 & .2 \\ .1 & .1 & .3 \\ .5 & .5 & .5 \end{bmatrix} \begin{bmatrix} x \\ y \\ z \end{bmatrix} = \begin{bmatrix} x \\ y \\ z \end{bmatrix} \end{cases}$

14. $A = \begin{array}{c} \\ A \\ B \end{array} \begin{array}{c} A B \\ \begin{bmatrix} .9 & .2 \\ .1 & .8 \end{bmatrix} \end{array}$

$A^2 \begin{bmatrix} \\ \end{bmatrix}_0 = \begin{bmatrix} .9 & .2 \\ .1 & .8 \end{bmatrix} \begin{bmatrix} .9 & .2 \\ .1 & .8 \end{bmatrix} \begin{bmatrix} .5 \\ .5 \end{bmatrix}_0 = \begin{bmatrix} .585 \\ .415 \end{bmatrix}$

58.5% of the regular listeners will listen to station A two days from now.

15. a. If the traffic is moderate on a particular day then for the next day the probability of light traffic is .2, the probability of moderate traffic is .75, and the probability of heavy traffic is .05.

b. Find the stable distribution.

$$\begin{cases} x + y + z = 1 \\ \begin{bmatrix} .70 & .20 & .10 \\ .20 & .75 & .30 \\ .10 & .05 & .60 \end{bmatrix} \begin{bmatrix} x \\ y \\ z \end{bmatrix} = \begin{bmatrix} x \\ y \\ z \end{bmatrix} \end{cases}$$

$$\begin{cases} x + y + z = 1 \\ .7x + .2y + .1z = x \\ .2x + .75y + .3z = y \\ .1x + .05y + .6z = z \end{cases}$$

$$\begin{cases} x + y + z = 1 \\ -.3x + .2y + .1z = 0 \\ .2x - .25y + .3z = 0 \\ .1x + .05y - .4z = 0 \end{cases}$$

$$\left[\begin{array}{ccc|c} 1 & 1 & 1 & 1 \\ -.3 & .2 & .1 & 0 \\ .2 & -.25 & .3 & 0 \\ .1 & .05 & -.4 & 0 \end{array}\right] \begin{array}{l} \\ 10[2] \\ 20[3] \\ 20[4] \end{array} \longrightarrow \left[\begin{array}{ccc|c} 1 & 1 & 1 & 1 \\ -3 & 2 & 1 & 0 \\ 4 & -5 & 6 & 0 \\ 2 & 1 & -8 & 0 \end{array}\right]$$

$$\begin{array}{l} [2]+3[1] \\ [3]+(-4)[1] \\ [4]+(-2)[1] \end{array} \longrightarrow \left[\begin{array}{ccc|c} 1 & 1 & 1 & 1 \\ 0 & 5 & 4 & 3 \\ 0 & -9 & 2 & -4 \\ 0 & -1 & -10 & -2 \end{array}\right]$$

$$\begin{array}{l} \frac{1}{5}[2] \\ [1]+(-1)[2] \\ [3]+9[2] \\ [4]+1[2] \end{array} \longrightarrow \left[\begin{array}{ccc|c} 1 & 0 & \frac{1}{5} & \frac{2}{5} \\ 0 & 1 & \frac{4}{5} & \frac{3}{5} \\ 0 & 0 & \frac{46}{5} & \frac{7}{5} \\ 0 & 0 & -\frac{46}{5} & -\frac{7}{5} \end{array}\right]$$

$$\begin{array}{l} \frac{5}{46}[3] \\ [1]+\left(-\frac{1}{5}\right)[3] \\ [2]+\left(-\frac{4}{5}\right)[3] \\ [4]+\frac{46}{5}[3] \end{array} \longrightarrow \left[\begin{array}{ccc|c} 1 & 0 & 0 & \frac{17}{46} \\ 0 & 1 & 0 & \frac{11}{23} \\ 0 & 0 & 1 & \frac{7}{46} \\ 0 & 0 & 0 & 0 \end{array}\right]$$

The stable distribution is $\begin{bmatrix} \frac{17}{46} \\ \frac{11}{23} \\ \frac{7}{46} \end{bmatrix}$ or about $\begin{bmatrix} .370 \\ .478 \\ .152 \end{bmatrix}$.

About 37.0% of workdays will have light traffic, 47.8% will have moderate traffic, and 15.2% will have heavy traffic.

c. $\dfrac{7}{46} \cdot 20 \approx 3.04$

About 3 workdays will have heavy traffic.

16.

$$\begin{array}{c} \\ C \\ L \\ G \\ S \end{array} \begin{array}{cccc} C & L & G & S \end{array}$$

$$\begin{array}{c} C \\ L \\ G \\ S \end{array} \left[\begin{array}{cc|cc} 1 & 0 & .60 & .05 \\ 0 & 1 & .10 & .40 \\ \hline 0 & 0 & .20 & .50 \\ 0 & 0 & .10 & .05 \end{array} \right]$$

$$R = \begin{bmatrix} .20 & .50 \\ .10 & .05 \end{bmatrix}$$

$$I - R = \begin{bmatrix} 1 & 0 \\ 0 & 1 \end{bmatrix} - \begin{bmatrix} .20 & .50 \\ .10 & .05 \end{bmatrix} = \begin{bmatrix} .80 & -.50 \\ -.10 & .95 \end{bmatrix} = \begin{bmatrix} a & b \\ c & d \end{bmatrix}$$

$$\Delta = ad - bc = (.80)(.95) - (-.50)(-.10) = .71$$

$$F = (I - R)^{-1} = \begin{bmatrix} \frac{d}{\Delta} & -\frac{b}{\Delta} \\ -\frac{c}{\Delta} & \frac{a}{\Delta} \end{bmatrix} = \begin{bmatrix} \frac{.95}{.71} & -\frac{.50}{.71} \\ -\frac{.10}{.71} & \frac{.80}{.71} \end{bmatrix} = \begin{array}{c} \\ G \\ S \end{array} \begin{array}{cc} G & S \\ \begin{bmatrix} \frac{95}{71} & \frac{50}{71} \\ \frac{10}{71} & \frac{80}{71} \end{bmatrix} \end{array}$$

Add the entries in each column:

$$\frac{95}{71} + \frac{10}{71} = \frac{105}{71} = 1\frac{34}{71} \approx 1.48$$

$$\frac{50}{71} + \frac{86}{71} = \frac{130}{71} = 1\frac{59}{71} \approx 1.83$$

A patient who begins in state G has an expected number of approximately 1.48 months; a patient who begins in state S has an expected number of approximately 1.83 months.

17. a.

$$\begin{cases} x + y + z + w = 1 \\ \begin{bmatrix} .20 & .10 & .05 & .05 \\ .30 & .20 & .20 & .30 \\ .40 & .40 & .50 & .40 \\ .10 & .30 & .25 & .25 \end{bmatrix} \begin{bmatrix} x \\ y \\ z \\ w \end{bmatrix} = \begin{bmatrix} x \\ y \\ z \\ w \end{bmatrix} \end{cases}$$

$$\begin{cases} x + y + z + w = 1 \\ .2x + .1y + .05z + .05w = x \\ .3x + .2y + .2z + .3w = y \\ .4x + .4y + .5z + .4w = z \\ .1x + .3y + .25z + .25w = w \end{cases}$$

$$\begin{cases} x + y + z + w = 1 \\ -.8x + .1y + .05z + .05w = 0 \\ .3x - .8y + .2z + .3w = 0 \\ .4x + .4y - .5z + .4w = 0 \\ .1x + .3y + .25z - .75w = 0 \end{cases}$$

$$\begin{bmatrix} 1 & 1 & 1 & 1 & | & 1 \\ -.8 & .1 & .05 & .05 & | & 0 \\ .3 & -.8 & .2 & .3 & | & 0 \\ .4 & .4 & -.5 & .4 & | & 0 \\ .1 & .3 & .25 & -.75 & | & 0 \end{bmatrix} \begin{matrix} \\ 20[2] \\ 10[3] \\ 10[4] \\ 20[5] \end{matrix} \longrightarrow \begin{bmatrix} 1 & 1 & 1 & 1 & | & 1 \\ -16 & 2 & 1 & 1 & | & 0 \\ 3 & -8 & 2 & 3 & | & 0 \\ 4 & 4 & -5 & 4 & | & 0 \\ 2 & 6 & 5 & -15 & | & 0 \end{bmatrix}$$

$$\begin{matrix} [2]+16[1] \\ [3]+(-3)[1] \\ [4]+(-4)[1] \\ [5]+(-2)[1] \end{matrix} \longrightarrow \begin{bmatrix} 1 & 1 & 1 & 1 & | & 1 \\ 0 & 18 & 17 & 17 & | & 16 \\ 0 & -11 & -1 & 0 & | & -3 \\ 0 & 0 & -9 & 0 & | & -4 \\ 0 & 4 & 3 & -17 & | & -2 \end{bmatrix}$$

$$\begin{matrix} \frac{1}{18}[2] \\ [1]+(-1)[2] \\ [3]+11[2] \\ [5]+(-4)[2] \end{matrix} \longrightarrow \begin{bmatrix} 1 & 0 & \frac{1}{18} & \frac{1}{18} & | & \frac{1}{9} \\ 0 & 1 & \frac{17}{18} & \frac{17}{18} & | & \frac{8}{9} \\ 0 & 0 & \frac{169}{18} & \frac{187}{18} & | & \frac{61}{9} \\ 0 & 0 & -9 & 0 & | & -4 \\ 0 & 0 & -\frac{7}{9} & -\frac{187}{9} & | & -\frac{50}{9} \end{bmatrix}$$

$$\begin{matrix} \frac{18}{169}[3] \\ [1]+\left(-\frac{1}{18}\right)[3] \\ [2]+\left(-\frac{17}{18}\right)[3] \\ [4]+9[3] \\ [5]+\frac{7}{9}[3] \end{matrix} \longrightarrow \begin{bmatrix} 1 & 0 & 0 & -\frac{1}{169} & | & \frac{12}{169} \\ 0 & 1 & 0 & -\frac{17}{169} & | & \frac{35}{169} \\ 0 & 0 & 1 & \frac{187}{169} & | & \frac{122}{169} \\ 0 & 0 & 0 & \frac{1683}{169} & | & \frac{422}{169} \\ 0 & 0 & 0 & -\frac{3366}{169} & | & -\frac{844}{169} \end{bmatrix}$$

$$\begin{matrix} \frac{169}{1683}[4] \\ [1]+\frac{1}{169}[4] \\ [2]+\frac{17}{169}[4] \\ [3]+\left(-\frac{187}{169}\right)[4] \\ [5]+\frac{3366}{169}[4] \end{matrix} \longrightarrow \begin{bmatrix} 1 & 0 & 0 & 0 & | & \frac{122}{1683} \\ 0 & 1 & 0 & 0 & | & \frac{23}{99} \\ 0 & 0 & 1 & 0 & | & \frac{4}{9} \\ 0 & 0 & 0 & 1 & | & \frac{422}{1683} \\ 0 & 0 & 0 & 0 & | & 0 \end{bmatrix}$$

The stable distribution is $\begin{bmatrix} \frac{122}{1683} \\ \frac{23}{99} \\ \frac{4}{9} \\ \frac{422}{1683} \end{bmatrix}$.

In the long run, the probability of having 1, 2, 3, or 4 units of water in the reservoir at any given time will be $\frac{122}{1683}, \frac{23}{99}, \frac{4}{9},$ or $\frac{422}{1683}$, respectively.

b. $\frac{122}{1683}(\$4000) + \frac{23}{99}(\$6000) + \frac{4}{9}(\$10,000) + \frac{422}{1683}(\$3000) \approx \$6881$

The average weekly benefits will be about $6881.

Chapter 9

Exercises 1

1. $R: \begin{bmatrix} -1 & \underline{-2} \\ \underline{0} & 3 \end{bmatrix}$, row 2; $C: \begin{bmatrix} -1 & -2 \\ \underline{0} & \underline{3} \end{bmatrix}$, column 1

3. $R: \begin{bmatrix} \underline{-2} & 4 & 1 \\ \underline{-1} & 3 & 5 \\ \underline{-3} & 5 & 2 \end{bmatrix}$, row 2;

$C: \begin{bmatrix} -2 & 4 & 1 \\ \underline{-1} & 3 & \underline{5} \\ -3 & \underline{5} & 2 \end{bmatrix}$, column 1

5. $R: \begin{bmatrix} \underline{0} & 3 \\ \underline{-1} & 1 \\ -2 & \underline{-4} \end{bmatrix}$, row 1;

$C: \begin{bmatrix} \underline{0} & 3 \\ -1 & 1 \\ -2 & -4 \end{bmatrix}$, column 1

7. Row minima: $\begin{bmatrix} 1 & \underline{0} \\ 0 & \underline{-1} \end{bmatrix}$,

column maxima: $\begin{bmatrix} \underline{1} & 0 \\ 0 & -1 \end{bmatrix}$

a. Row 1, column 2

b. 0

9. $\begin{array}{c} \\ H \\ T \end{array} \begin{array}{cc} H & T \\ \begin{bmatrix} 2 & -1 \\ -1 & -4 \end{bmatrix} \end{array}$

Row minima: $\begin{bmatrix} 2 & \underline{-1} \\ -1 & \underline{-4} \end{bmatrix}$,

column maxima: $\begin{bmatrix} \underline{2} & -1 \\ -1 & -4 \end{bmatrix}$

Row 1, column 2 is a saddle point, so the game is strictly determined. R should show heads, C should show tails.

11. $\begin{array}{c} \\ F \\ A \\ N \end{array} \begin{array}{ccc} F & A & N \\ \begin{bmatrix} 8000 & -1000 & 1000 \\ -7000 & 4000 & -2000 \\ 3000 & 3000 & 2000 \end{bmatrix} \end{array}$

Row minima: $\begin{bmatrix} 1000 & \underline{-1000} & 1000 \\ \underline{-7000} & 4000 & -2000 \\ 3000 & 3000 & \underline{2000} \end{bmatrix}$

Column maxima: $\begin{bmatrix} \underline{8000} & -1000 & 1000 \\ -7000 & \underline{4000} & -2000 \\ 3000 & 3000 & \underline{2000} \end{bmatrix}$

Row 3, column 3 is a saddle point, so the game is strictly determined. Both candidates should be neutral.

13. $\begin{array}{c} \\ 5 \\ 10 \end{array} \begin{array}{ccc} 6 & 7 & 8 \\ \begin{bmatrix} 1 & -5 & -5 \\ -6 & 3 & 2 \end{bmatrix} \end{array}$

Row minima: $\begin{bmatrix} 1 & -5 & \underline{-5} \\ \underline{-6} & 3 & 2 \end{bmatrix}$

Column maxima: $\begin{bmatrix} \underline{1} & -5 & -5 \\ -6 & \underline{3} & \underline{2} \end{bmatrix}$

No saddle point, so the game is not strictly determined.

Exercises 2

1. a. $[.5 \quad .5] \begin{bmatrix} 3 & -1 \\ -7 & 5 \end{bmatrix} \begin{bmatrix} .5 \\ .5 \end{bmatrix} = [0]$

b. $[1 \quad 0] \begin{bmatrix} 3 & -1 \\ -7 & 5 \end{bmatrix} \begin{bmatrix} .5 \\ .5 \end{bmatrix} = [1]$

c. $[.3 \quad .7] \begin{bmatrix} 3 & -1 \\ -7 & 5 \end{bmatrix} \begin{bmatrix} .6 \\ .4 \end{bmatrix} = [-1.12]$

d. $[.75 \quad .25] \begin{bmatrix} 3 & -1 \\ -7 & 5 \end{bmatrix} \begin{bmatrix} .2 \\ .8 \end{bmatrix} = [.5]$

(b) is most advantageous to R.

3. $[.3 \quad .7]\begin{bmatrix} -20,000 & 0 \\ 0 & -50,000 \end{bmatrix}\begin{bmatrix} .2 \\ .8 \end{bmatrix} = [-29,200]$

$29,200

5. The payoff matrix is $\begin{matrix} & V & C \\ V & \begin{bmatrix} 0 & 2 \\ C & -1 & 0 \end{bmatrix} \end{matrix}$.

$[.25 \quad .75]\begin{bmatrix} 0 & 2 \\ -1 & 0 \end{bmatrix}\begin{bmatrix} .4 \\ .6 \end{bmatrix} = [0]$

Zero

Exercises 3

1. Maximize $M = z_1 + z_2$ subject to

$$\begin{cases} 2z_1 + 4z_2 \le 1 \\ 5z_1 + 3z_2 \le 1 \\ z_1 \ge 0, \ z_2 \ge 0 \end{cases}$$

$$\begin{array}{c|ccccc|c} & z_1 & z_2 & t & u & M & \\ \hline t & 2 & 4 & 1 & 0 & 0 & 1 \\ u & 5 & 3 & 0 & 1 & 0 & 1 \\ M & -1 & -1 & 0 & 0 & 1 & 0 \end{array}$$

$$\begin{array}{c|ccccc|c} & z_1 & z_2 & t & u & M & \\ \hline t & 0 & \frac{14}{5} & 1 & -\frac{2}{5} & 0 & \frac{3}{5} \\ z_1 & 1 & \frac{3}{5} & 0 & \frac{1}{5} & 0 & \frac{1}{5} \\ M & 0 & -\frac{2}{5} & 0 & \frac{1}{5} & 1 & \frac{1}{5} \end{array}$$

$$\begin{array}{c|ccccc|c} & z_1 & z_2 & t & u & M & \\ \hline z_2 & 0 & 1 & \frac{5}{14} & -\frac{1}{7} & 0 & \frac{3}{14} \\ z_1 & 1 & 0 & -\frac{3}{14} & \frac{2}{7} & 0 & \frac{1}{14} \\ M & 0 & 0 & \frac{1}{7} & \frac{1}{7} & 1 & \frac{2}{7} \end{array}$$

$z_1 = \dfrac{3}{14}$, $z_2 = \dfrac{1}{14}$, $M = \dfrac{2}{7}$, $v = \dfrac{1}{M} = \dfrac{7}{2}$, and

the optimal strategy for C is $\begin{bmatrix} vz_1 \\ vz_2 \end{bmatrix} = \begin{bmatrix} \frac{1}{4} \\ \frac{3}{4} \end{bmatrix}$.

The optimal strategy for R is given by the bottom entries under t and u:

$[vt \quad vu] = \left[\frac{7}{2} \cdot \frac{1}{7} \quad \frac{7}{2} \cdot \frac{1}{7}\right] = \left[\frac{1}{2} \quad \frac{1}{2}\right]$.

3. Add 7 to each entry to make all the entries positive. We get $\begin{bmatrix} 10 & 1 \\ 2 & 11 \end{bmatrix}$. Then maximize

$M = z_1 + z_2$ subject to $\begin{cases} 10z_1 + z_2 \le 1 \\ 2z_1 + 11z_2 \le 1 \\ z_1 \ge 0, \ z_2 \ge 0 \end{cases}$

$$\begin{array}{c|ccccc|c} & z_1 & z_2 & t & u & M & \\ \hline t & 10 & 1 & 1 & 0 & 0 & 1 \\ u & 2 & 11 & 0 & 1 & 0 & 1 \\ M & -1 & -1 & 0 & 0 & 1 & 0 \end{array}$$

$$\begin{array}{c|ccccc|c} & z_1 & z_2 & t & u & M & \\ \hline z_1 & 1 & \frac{1}{10} & \frac{1}{10} & 0 & 0 & \frac{1}{10} \\ u & 0 & \frac{54}{5} & -\frac{1}{5} & 1 & 0 & \frac{4}{5} \\ M & 0 & -\frac{9}{10} & \frac{1}{10} & 0 & 1 & \frac{1}{10} \end{array}$$

$$\begin{array}{c|ccccc|c} & z_1 & z_2 & t & u & M & \\ \hline z_1 & 1 & 0 & \frac{11}{108} & -\frac{1}{108} & 0 & \frac{5}{54} \\ z_2 & 0 & 1 & -\frac{1}{54} & \frac{5}{54} & 0 & \frac{2}{27} \\ M & 0 & 0 & \frac{1}{12} & \frac{1}{12} & 1 & \frac{1}{6} \end{array}$$

$z_1 = \dfrac{5}{54}$, $z_2 = \dfrac{2}{27}$, $M = \dfrac{1}{6}$, $v = \dfrac{1}{M} = 6$, and

the optimal strategy for C is $\begin{bmatrix} vz_1 \\ vz_2 \end{bmatrix} = \begin{bmatrix} \frac{5}{9} \\ \frac{4}{9} \end{bmatrix}$.

The optimal strategy for R is given by the bottom entries under t and u:

$[vt \quad vu] = \left[6 \cdot \frac{1}{12} \quad 6 \cdot \frac{1}{12}\right] = \left[\frac{1}{2} \quad \frac{1}{2}\right]$

5. Maximize $M = z_1 + z_2$ subject to

$$\begin{cases} 4z_1 + z_2 \le 1 \\ 2z_1 + 4z_2 \le 1 \\ z_1 \ge 0, \ z_2 \ge 0 \end{cases}$$

$$\begin{array}{c|ccccc|c} & z_1 & z_2 & t & u & M & \\ \hline t & 4 & 1 & 1 & 0 & 0 & 1 \\ u & 2 & 4 & 0 & 1 & 0 & 1 \\ M & -1 & -1 & 0 & 0 & 1 & 0 \end{array}$$

$$\begin{array}{c|ccccc|c} & z_1 & z_2 & t & u & M & \\ \hline t & \frac{7}{2} & 0 & 1 & -\frac{1}{4} & 0 & \frac{3}{4} \\ z_2 & \frac{1}{2} & 1 & 0 & \frac{1}{4} & 0 & \frac{1}{4} \\ \hline M & -\frac{1}{2} & 0 & 0 & \frac{1}{4} & 1 & \frac{1}{4} \end{array}$$

$$\begin{array}{c|ccccc|c} & z_1 & z_2 & t & u & M & \\ \hline z_1 & 1 & 0 & \frac{2}{7} & -\frac{1}{14} & 0 & \frac{3}{14} \\ z_2 & 0 & 1 & -\frac{1}{7} & \frac{2}{7} & 0 & \frac{1}{7} \\ \hline M & 0 & 0 & \frac{1}{7} & \frac{3}{14} & 1 & \frac{5}{14} \end{array}$$

$z_1 = \dfrac{3}{14}$, $z_2 = \dfrac{1}{7}$, $M = \dfrac{5}{14}$, $v = \dfrac{1}{M} = \dfrac{14}{5}$, and

the optimal strategy for C is $\begin{bmatrix} vz_1 \\ vz_2 \end{bmatrix} = \begin{bmatrix} \frac{3}{5} \\ \frac{2}{5} \end{bmatrix}$.

The optimal strategy for R is given by the bottom entries under t and u:

$[vt \quad vu] = \left[\frac{14}{5} \cdot \frac{1}{7} \quad \frac{14}{5} \cdot \frac{3}{14} \right] = \left[\frac{2}{5} \quad \frac{3}{5} \right]$

7. Add 2 to each entry to make all entries of the payoff matrix.

$$\begin{bmatrix} 5 & 7 & 1 \\ 6 & 1 & 8 \end{bmatrix}$$

Set the tableaux up to find C's optimal strategy, then read the dual's solution off the bottom row:

$$\begin{array}{c|cccccc|c} & z_1 & z_2 & z_3 & t & u & M & \\ \hline t & 5 & 7 & 1 & 1 & 0 & 0 & 1 \\ u & 6 & 1 & 7 & 0 & 1 & 0 & 1 \\ \hline M & -1 & -1 & -1 & 0 & 0 & 1 & 0 \end{array}$$

$$\begin{array}{c|cccccc|c} & z_1 & z_2 & z_3 & t & u & M & \\ \hline z_2 & \frac{5}{7} & 1 & \frac{1}{7} & \frac{1}{7} & 0 & 0 & \frac{1}{7} \\ u & \frac{37}{7} & 0 & \frac{55}{7} & -\frac{1}{7} & 1 & 0 & \frac{6}{7} \\ \hline M & -\frac{2}{7} & 0 & -\frac{6}{7} & \frac{1}{7} & 0 & 1 & \frac{1}{7} \end{array}$$

$$\begin{array}{c|cccccc|c} & z_1 & z_2 & z_3 & t & u & M & \\ \hline z_2 & \frac{34}{55} & 1 & 0 & \frac{8}{55} & -\frac{1}{55} & 0 & \frac{7}{55} \\ z_3 & \frac{37}{55} & 0 & 1 & -\frac{1}{55} & \frac{7}{55} & 0 & \frac{6}{55} \\ \hline M & \frac{16}{55} & 0 & 0 & \frac{7}{55} & \frac{6}{55} & 1 & \frac{13}{55} \end{array}$$

$t = \dfrac{7}{55}$, $u = \dfrac{6}{55}$, $M = \dfrac{13}{55}$, $v = \dfrac{1}{M} = \dfrac{55}{13}$, and

R's optimal strategy is $[vt \quad vu] = \left[\frac{7}{13} \quad \frac{6}{13} \right]$.

9. Row 2, column 2 is a saddle point. $\begin{bmatrix} 0 \\ 1 \end{bmatrix}$

11. The payoff matrix, with entries in thousands of dollars, is $\begin{array}{c} \\ 1 \\ 2 \end{array}\begin{array}{cc} 1 & 2 \\ \end{array}\begin{bmatrix} -2 & 7 \\ 7 & -1 \end{bmatrix}$. Add 3 to each entry to make all the entries positive. We get $\begin{bmatrix} 1 & 10 \\ 10 & 2 \end{bmatrix}$. Then maximize $M = z_1 + z_2$

subject to $\begin{cases} z_1 + 10z_2 \le 1 \\ 10z_1 + 2z_2 \le 1. \\ z_1 \ge 0,\ z_2 \ge 0 \end{cases}$

$$\begin{array}{c|ccccc|c} & z_1 & z_2 & t & u & M & \\ \hline t & 1 & \underline{10} & 1 & 0 & 0 & 1 \\ u & 10 & 2 & 0 & 1 & 0 & 1 \\ \hline M & -1 & -1 & 0 & 0 & 1 & 0 \end{array}$$

$$\begin{array}{c|ccccc|c} & z_1 & z_2 & t & u & M & \\ \hline z_2 & \frac{1}{10} & 1 & \frac{1}{10} & 0 & 0 & \frac{1}{10} \\ u & \frac{49}{5} & 0 & -\frac{1}{5} & 1 & 0 & \frac{4}{5} \\ \hline M & -\frac{9}{10} & 0 & \frac{1}{10} & 0 & 1 & \frac{1}{10} \end{array}$$

$$\begin{array}{c|ccccc|c} & z_1 & z_2 & t & u & M & \\ \hline z_2 & 0 & 1 & \frac{5}{49} & -\frac{1}{98} & 0 & \frac{9}{98} \\ z_1 & 1 & 0 & -\frac{5}{49} & \frac{5}{49} & 0 & \frac{4}{49} \\ \hline M & 0 & 0 & \frac{4}{49} & \frac{9}{98} & 1 & \frac{17}{98} \end{array}$$

$v = \dfrac{1}{M} = \dfrac{98}{17}$

a. R's optimal strategy is given by the bottom entries under t and u:

$[vt \quad vu] = \left[\frac{8}{17} \quad \frac{9}{17} \right]$

b. C's optimal strategy is $\begin{bmatrix} vz_1 \\ vz_2 \end{bmatrix} = \begin{bmatrix} \frac{8}{17} \\ \frac{9}{17} \end{bmatrix}$.

c. $v - 3 = \dfrac{1}{M} - 3 = \dfrac{98}{17} - 3 \approx 2.765$

The value of the game is about \$2765.

13. Add 4 to each entry to make all the entries positive, then maximize $M = z_1 + z_2$ subject

to $\begin{cases} 2z_1 + 5z_2 \le 1 \\ 6z_1 + z_2 \le 1 \\ 5z_1 + z_2 \le 1 \\ z_1 \ge 0,\ z_2 \ge 0 \end{cases}$.

$$
\begin{array}{c}
 \\ s \\ t \\ u \\ M
\end{array}
\begin{array}{c}
\begin{array}{cccccc} z_1 & z_2 & s & t & u & M \end{array} \\
\left[\begin{array}{cccccc|c}
2 & \underline{5} & 1 & 0 & 0 & 0 & 1 \\
6 & 1 & 0 & 1 & 0 & 0 & 1 \\
5 & 2 & 0 & 0 & 1 & 0 & 1 \\
\hline
-1 & -1 & 0 & 0 & 0 & 1 & 0
\end{array}\right]
\end{array}
$$

$$
\begin{array}{c}
 \\ z_2 \\ t \\ u \\ M
\end{array}
\begin{array}{c}
\begin{array}{cccccc} z_1 & z_2 & s & t & u & M \end{array} \\
\left[\begin{array}{cccccc|c}
\frac{2}{5} & 1 & \frac{1}{5} & 0 & 0 & 0 & \frac{1}{5} \\
\frac{28}{5} & 0 & -\frac{1}{5} & 1 & 0 & 0 & \frac{4}{5} \\
\frac{21}{5} & 0 & -\frac{2}{5} & 0 & 1 & 0 & \frac{3}{5} \\
\hline
-\frac{3}{5} & 0 & \frac{1}{5} & 0 & 0 & 1 & \frac{1}{5}
\end{array}\right]
\end{array}
$$

$$
\begin{array}{c}
 \\ z_2 \\ z_1 \\ u \\ M
\end{array}
\begin{array}{c}
\begin{array}{cccccc} z_1 & z_2 & s & t & u & M \end{array} \\
\left[\begin{array}{cccccc|c}
0 & 1 & \frac{3}{14} & -\frac{1}{14} & 0 & 0 & \frac{1}{7} \\
1 & 0 & -\frac{1}{28} & \frac{5}{28} & 0 & 0 & \frac{1}{7} \\
0 & 0 & -\frac{1}{4} & -\frac{3}{4} & 1 & 0 & 0 \\
\hline
0 & 0 & \frac{5}{28} & \frac{3}{28} & 0 & 1 & \frac{2}{7}
\end{array}\right]
\end{array}
$$

$$v = \frac{1}{M} = \frac{7}{2}$$

C's optimal strategy is $\begin{bmatrix} vz_1 \\ vz_2 \end{bmatrix} = \begin{bmatrix} \frac{1}{2} \\ \frac{1}{2} \end{bmatrix}$. R's

optimal strategy is given by the bottom entries under s, t, and u:

$$[vs \quad vt \quad vu] = \begin{bmatrix} \frac{5}{8} & \frac{3}{8} & 0 \end{bmatrix}$$

Chapter 9 Supplementary Exercises

1. Row minima: $\begin{bmatrix} 5 & \underline{-1} & 1 \\ \underline{-3} & 5 & 1 \\ 4 & 3 & \underline{2} \end{bmatrix}$,

column maxima: $\begin{bmatrix} \underline{5} & -1 & 1 \\ -3 & \underline{5} & 1 \\ 4 & 3 & \underline{2} \end{bmatrix}$

The game is strictly determined, with a saddle point at row 3, column 3 and a value of 2.

2. Row minima: $\begin{bmatrix} 1 & 2 & 3 \\ 3 & 2 & \underline{1} \end{bmatrix}$,

column maxima: $\begin{bmatrix} 1 & 2 & 3 \\ \underline{3} & \underline{2} & 1 \end{bmatrix}$

The game has no saddle point and so is not strictly determined.

3. Row minima: $\begin{bmatrix} \underline{0} & 1 \\ 1 & \underline{0} \\ 2 & \underline{-1} \end{bmatrix}$,

column maxima: $\begin{bmatrix} 0 & 1 \\ 1 & 0 \\ \underline{2} & -1 \end{bmatrix}$

The game has no saddle point and so is not strictly determined.

4. Row minima: $\begin{bmatrix} 2 & \underline{1} & 2 \\ \underline{-1} & 0 & 3 \\ 4 & 1 & \underline{-4} \end{bmatrix}$,

column maxima: $\begin{bmatrix} 2 & 1 & 2 \\ -1 & 0 & \underline{3} \\ \underline{4} & 1 & -4 \end{bmatrix}$

The game is strictly determined, with a saddle point at row 1, column 2 and a value of 1.

5. $\begin{bmatrix} \frac{3}{4} & \frac{1}{4} \end{bmatrix} \begin{bmatrix} 0 & 24 \\ 12 & -36 \end{bmatrix} \begin{bmatrix} \frac{1}{3} \\ \frac{2}{3} \end{bmatrix} = [7]$

7

6. $\begin{bmatrix} \frac{1}{2} & \frac{1}{2} \end{bmatrix} \begin{bmatrix} -6 & 6 & 0 \\ 0 & -12 & 24 \end{bmatrix} \begin{bmatrix} \frac{1}{3} \\ \frac{1}{3} \\ \frac{1}{3} \end{bmatrix} = [2]$

2

7. $[.2 \quad .3 \quad .5] \begin{bmatrix} 1 & 0 \\ -3 & 1 \\ 0 & 5 \end{bmatrix} \begin{bmatrix} .4 \\ .6 \end{bmatrix} = [1.4]$

1.4

8. $[.1 \quad .1 \quad .8]\begin{bmatrix} 0 & 1 & 3 \\ -1 & 0 & 2 \\ -3 & -2 & 0 \end{bmatrix}\begin{bmatrix} .4 \\ .3 \\ .3 \end{bmatrix} = [-1.3]$

-1.3

9. Add 4 to each entry to get $\begin{bmatrix} 1 & 8 \\ 6 & 2 \end{bmatrix}$. Then use

the simplex method.

$$\begin{array}{c} \\ t \\ u \\ M \end{array}\begin{bmatrix} z_1 & z_2 & t & u & M & \\ 1 & 8 & 1 & 0 & 0 & 1 \\ \underline{6} & 2 & 0 & 1 & 0 & 1 \\ -1 & -1 & 0 & 0 & 1 & 0 \end{bmatrix}$$

$$\begin{array}{c} \\ t \\ z_1 \\ M \end{array}\begin{bmatrix} z_1 & z_2 & t & u & M & \\ 0 & \frac{23}{3} & 1 & -\frac{1}{6} & 0 & \frac{5}{7} \\ 1 & \frac{1}{3} & 0 & \frac{1}{6} & 0 & \frac{1}{6} \\ 0 & -\frac{2}{3} & 0 & \frac{1}{6} & 1 & \frac{1}{6} \end{bmatrix}$$

$$\begin{array}{c} \\ z_2 \\ z_1 \\ M \end{array}\begin{bmatrix} z_1 & z_2 & t & u & M & \\ 0 & 1 & \frac{3}{23} & -\frac{1}{46} & 0 & \frac{5}{46} \\ 1 & 0 & -\frac{1}{23} & \frac{4}{23} & 0 & \frac{3}{23} \\ 0 & 0 & \frac{2}{23} & \frac{7}{46} & 1 & \frac{11}{46} \end{bmatrix}$$

$v = \dfrac{1}{M} = \dfrac{46}{11}$

R's optimal strategy:

$[vt \quad vu] = \left[\frac{46}{11} \cdot \frac{2}{23} \quad \frac{46}{11} \cdot \frac{7}{46}\right] = \left[\frac{4}{11} \quad \frac{7}{11}\right]$

C's optimal strategy:

$\begin{bmatrix} vz_1 \\ vz_2 \end{bmatrix} = \begin{bmatrix} \frac{46}{11} \cdot \frac{3}{46} \\ \frac{46}{11} \cdot \frac{5}{23} \end{bmatrix} = \begin{bmatrix} \frac{5}{11} \\ \frac{6}{11} \end{bmatrix}$

10. Add 7 to each entry to get $\begin{bmatrix} 10 & 1 \\ 8 & 11 \end{bmatrix}$. Then

apply the simplex method.

$$\begin{array}{c} \\ \\ M \end{array}\begin{bmatrix} z_1 & z_2 & t & u & M & \\ \underline{10} & 1 & 1 & 0 & 0 & 1 \\ 3 & 11 & 0 & 1 & 0 & 1 \\ -1 & -1 & 0 & 0 & 1 & 0 \end{bmatrix}$$

$$\begin{array}{c} \\ z_1 \\ u \\ M \end{array}\begin{bmatrix} z_1 & z_2 & t & u & M & \\ 1 & \frac{1}{10} & \frac{1}{10} & 0 & 0 & \frac{1}{10} \\ 0 & \frac{107}{10} & -\frac{3}{10} & 1 & 0 & \frac{7}{10} \\ 0 & -\frac{9}{10} & \frac{1}{10} & 0 & 1 & \frac{1}{10} \end{bmatrix}$$

$$\begin{array}{c} \\ z_1 \\ z_2 \\ M \end{array}\begin{bmatrix} z_1 & z_2 & t & u & M & \\ 1 & 0 & \frac{11}{107} & -\frac{1}{107} & 0 & \frac{10}{107} \\ 0 & 1 & -\frac{3}{107} & \frac{10}{107} & 0 & \frac{7}{107} \\ 0 & 0 & \frac{8}{107} & \frac{9}{107} & 1 & \frac{17}{107} \end{bmatrix}$$

$v = \dfrac{1}{M} = \dfrac{107}{17}$

R's optimal strategy:

$[vt \quad vu] = \left[\frac{107}{17} \cdot \frac{8}{107} \quad \frac{107}{17} \cdot \frac{9}{107}\right] = \left[\frac{8}{17} \quad \frac{9}{17}\right]$

C's optimal strategy:

$\begin{bmatrix} vz_1 \\ vz_2 \end{bmatrix} = \begin{bmatrix} \frac{107}{17} \cdot \frac{10}{107} \\ \frac{107}{17} \cdot \frac{7}{107} \end{bmatrix} = \begin{bmatrix} \frac{10}{17} \\ \frac{7}{17} \end{bmatrix}$

11. Row 2, column 3 is a saddle point: $[0 \quad 1]$

12.
$$\begin{array}{c} \\ s \\ t \\ u \\ M \end{array}\begin{bmatrix} z_1 & z_2 & s & t & u & M & \\ 1 & 3 & 1 & 0 & 0 & 0 & 1 \\ 3 & 1 & 0 & 1 & 0 & 0 & 1 \\ \underline{4} & 2 & 0 & 0 & 1 & 0 & 1 \\ -1 & -1 & 0 & 0 & 0 & 1 & 0 \end{bmatrix}$$

$$\begin{array}{c} \\ s \\ t \\ z_1 \\ M \end{array}\begin{bmatrix} z_1 & z_2 & s & t & u & M & \\ 0 & \frac{5}{2} & 1 & 0 & -\frac{1}{4} & 0 & \frac{3}{4} \\ 0 & -\frac{1}{2} & 0 & 1 & -\frac{3}{4} & 0 & \frac{1}{4} \\ 1 & \frac{1}{2} & 0 & 0 & \frac{1}{4} & 0 & \frac{1}{4} \\ 0 & -\frac{1}{2} & 0 & 0 & \frac{1}{4} & 1 & \frac{1}{4} \end{bmatrix}$$

$$\begin{array}{c} \\ z_2 \\ t \\ z_1 \\ M \end{array}\begin{bmatrix} z_1 & z_2 & s & t & u & M & \\ 0 & 1 & \frac{2}{5} & 0 & -\frac{1}{10} & 0 & \frac{3}{10} \\ 0 & 0 & \frac{1}{5} & 1 & -\frac{4}{5} & 0 & \frac{2}{5} \\ 1 & 0 & -\frac{1}{5} & 0 & \frac{3}{10} & 0 & \frac{1}{10} \\ 0 & 0 & \frac{1}{5} & 0 & \frac{1}{5} & 1 & \frac{2}{5} \end{bmatrix}$$

$v = \dfrac{1}{M} = \dfrac{5}{2}$

$\begin{bmatrix} vz_1 \\ vz_2 \end{bmatrix} = \begin{bmatrix} \frac{5}{2} \cdot \frac{1}{10} \\ \frac{5}{2} \cdot \frac{3}{10} \end{bmatrix} = \begin{bmatrix} \frac{1}{4} \\ \frac{3}{4} \end{bmatrix}$

13. The payoff matrix is $\begin{array}{c} 2 \\ 6 \end{array}\begin{array}{c} 2 \quad\quad 6 \\ \begin{bmatrix} -3 & 2 \\ 6 & -3 \end{bmatrix} \end{array}$. Add 4 to

each entry to get $\begin{bmatrix} 1 & 6 \\ 10 & 1 \end{bmatrix}$. Then apply the

simplex method.

$$\begin{array}{c} \\ t \\ u \\ M \end{array}\begin{array}{ccccc} z_1 & z_2 & t & u & M \\ \end{array}\left[\begin{array}{ccccc|c} 1 & \underline{6} & 1 & 0 & 0 & 1 \\ 10 & 1 & 0 & 1 & 0 & 1 \\ -1 & -1 & 0 & 0 & 1 & 0 \end{array}\right]$$

$$\begin{array}{c} z_2 \\ u \\ M \end{array}\begin{array}{ccccc} z_1 & z_2 & t & u & M \\ \end{array}\left[\begin{array}{ccccc|c} \frac{1}{6} & 1 & \frac{1}{6} & 0 & 0 & \frac{1}{6} \\ \frac{59}{6} & 0 & -\frac{1}{6} & 1 & 0 & \frac{5}{6} \\ -\frac{5}{6} & 0 & \frac{1}{6} & 0 & 1 & \frac{1}{6} \end{array}\right]$$

$$\begin{array}{c} z_2 \\ z_1 \\ M \end{array}\begin{array}{ccccc} z_1 & z_2 & t & u & M \\ \end{array}\left[\begin{array}{ccccc|c} 0 & 1 & \frac{10}{59} & -\frac{1}{59} & 0 & \frac{9}{59} \\ 1 & 0 & -\frac{1}{59} & \frac{6}{59} & 0 & \frac{5}{59} \\ 0 & 0 & \frac{9}{59} & \frac{5}{59} & 1 & \frac{14}{59} \end{array}\right]$$

$$v = \frac{1}{M} = \frac{59}{14}$$

a. Carol's optimal strategy is
$$\begin{bmatrix} \frac{59}{14} \cdot \frac{5}{59} \\ \frac{59}{14} \cdot \frac{9}{59} \end{bmatrix} = \begin{bmatrix} \frac{5}{14} \\ \frac{9}{14} \end{bmatrix}.$$
Ruth's optimal strategy is
$$\begin{bmatrix} \frac{59}{14} \cdot \frac{9}{59} & \frac{59}{14} \cdot \frac{5}{59} \end{bmatrix} = \begin{bmatrix} \frac{9}{14} & \frac{5}{14} \end{bmatrix}.$$

b. Since the value is positive, the game favors Ruth (the row player). The value of the game is $\dfrac{59}{14} - 4 = \dfrac{3}{14}$.

14. a.

$$\begin{array}{c} \\ A \\ B \\ C \end{array}\begin{array}{ccc} \text{Strong} & \text{Avg.} & \text{Weak} \\ \end{array}\left[\begin{array}{ccc} 3000 & 2000 & 1000 \\ 6000 & 2000 & -3000 \\ 15{,}000 & 1000 & -10{,}000 \end{array}\right]$$

b. Row 1, column 3 is a saddle point. The investors optimal strategy is to buy stock *A*.

Chapter 10

Exercises 1

1. a. $i = \dfrac{.12}{12} = .01$
$n = 12 \times 2 = 24$

   b. $i = \dfrac{.08}{4} = .02$
$n = 4 \times 5 = 20$

   c. $i = \dfrac{.10}{2} = .05$
$n = 2 \times 20 = 40$

3. a. $i = \dfrac{.06}{1} = .06$
$n = 1 \times 4 = 4$
$P = \$500$
$F = (1 + .06)^4 \cdot 500 = \$631.24$

   b. $i = \dfrac{.06}{12} = .005$
$n = 12 \times 10 = 120$
$P = \$800$
$F = (1 + .005)^{120} \cdot 800 = \$1455.52$

   c. $i = \dfrac{.04}{2} = .02$
$n = 2 \times 9.5 = 19$
$P = \dfrac{1}{(1 + .02)^{19}} \cdot 9000 = \$6177.88$
$F = \$9000$

5. $\left(1 + \dfrac{.06}{12}\right)^{12 \times 2} (\$1000) = \$1127.16$

7. $\left[\dfrac{1}{\left(1 + \frac{.06}{12}\right)^{12 \times 25}}\right](\$100,000) = \$22,396.57$

9. $\left(1 + \dfrac{.06}{12}\right)^{12 \times 3} (\$6000) = \$7180.08$
$\$7180.08 - \$6000 = \$1180.08$

11. a. $\left[\dfrac{1}{\left(1+\frac{.06}{12}\right)^{12\times2}}\right]($4000) = 3548.74

b. $4000 - $3548.74 = 451.26

c.

Month	Interest	Balance
0		$3548.74
1	$17.74	$(1.005)($3548.74) = 3566.48
2	$17.84	$(1.005)^2($3548.74) = 3584.32
3	$17.92	$(1.005)^3($3548.74) = 3602.24

13. $\left(1+\dfrac{.04}{4}\right)^{4\times6.25}($10,000) = $12,824.32$

15. $\left[\dfrac{1}{\left(1+\frac{.04}{4}\right)^{4\times3}}\right]($10,000) = 8874.49

17. a. $\dfrac{r}{12}($1000.00) = 5.00, so $r = .06 = 6\%$

b. $(1.005)^3($1000.00) = 1015.08
$1015.08 - $1010.03 = 5.05

c. $(1.005)^{24}($1000.00) = 1127.16
$[(1.005)^{24} - (1.005)^{23}]($1000.00) = 5.61

19. For $P = 1000,
$$F = \left(1+\dfrac{.06}{1}\right)^9($1000) = $1689.48.$$
$1700 in 9 years is more profitable.

21. $\left(1+\dfrac{.26}{52}\right)^{52} \approx 1.296$, which is a 29.6%
increase in 1 year. This interest rate is worse.

23. $\dfrac{r}{4}($10,000) = 100, so $r = .04$ and $i = .01$.
$$\dfrac{1}{(1.01)^{12}}($10,000) = $8874.49$$

25. a. $r = .04$
$n = \dfrac{6}{12} = \dfrac{1}{2}$
$P = 500
$A = 510

b. $r = .05$
$n = 2$
$P = 500
$A = 550

27. $(1 + 3 \times .05)($1000) = 1150

29. $\left[\dfrac{1}{(1+2\times.10)}\right](\$3000) = \$2500$

31. $\left(1+\dfrac{6}{12}r\right)(\$980) = \$1000;$
$r \approx .0408 = 4.08\%$

33. $(1 + n \times .05)P = 2P;\ n = 20$ years

35. $A = (1 + nr)P;\ P = \dfrac{A}{1+nr}$

37. (a)

39. $\left(1+\dfrac{.04}{4}\right)^{4\times1}(\$100) = \$104.06$
$\dfrac{\$4.06}{\$100} = .0406 = 4.06\%$

41. $\left(1+\dfrac{.04}{2}\right)^{2\times1} = 1.0404;\ 4.04\%$

43. $\left(1+\dfrac{.06}{12}\right)^{12\times1} \approx 1.0617;\ 6.17\%$

45. $\left(1+\dfrac{r}{2}\right)^{2\times1} - 1 = r + \dfrac{r^2}{4}$

47. $\left(1+\dfrac{r}{n}\right)^{n} - 1$

49. After 5 months:
$(1.005)^5(\$10,000) = \$10,252.51$
After 10 months:
$(1.005)^{10}(\$10,000) = \$10,511.40$
After 15 months:
$(1.005)^{15}(\$10,000) = \$10,776.83$
After 30 months:
$(1.005)^{30}(\$10,000) = \$11,614.00$
After 53 months:
$(1.005)^{53}(\$10,000) = \$13,025.71$
$\$10,252.51; \$10,511.40; \$10,776.83; 30; 53$

51. $\left(1+\dfrac{.06}{4}\right)^{4n}(\$100)$ passes $200 when
$n = 11.75$ years.

Exercises 2

1. a. $i = \dfrac{.06}{12} = .005$
$n = 12 \times 10 = 120$
$R = \$50$
$F = \$8193.97$

    b. $i = \dfrac{.04}{2} = .02$
$n = 2 \times 10 = 20$
$R = \$2675.19$
$F = \$65,000$

3. $\left[\dfrac{\left(1+\frac{.06}{12}\right)^{12\times5} - 1}{\frac{.06}{12}}\right](\$100) = \$6977$

5. $\left[\dfrac{\frac{.08}{4}\left(1+\frac{.08}{4}\right)^{4\times7}}{\left(1+\frac{.08}{4}\right)^{4\times7} - 1}\right](\$100,000) = \$4698.97$

7. a. $\left[\dfrac{\left(1+\frac{.06}{12}\right)^{12\times4} - 1}{\frac{.06}{12}}\right](\$500) = \$27,048.92$

    b. $\$27,048.92 - 48(\$500) = \$3048.92$

    c.

Month	Interest	Balance
1		$500
2	2.50	$1002.50
3	5.01	$1507.51

9. $\left[\dfrac{\frac{.06}{12}}{\left(1+\frac{.06}{12}\right)^{12\times3} - 1}\right](\$12,000) = \$305.06$
Deposited: $36(\$305.06) = \$10,982.16$
Interest: $\$12,000 - \$10,982.16 = \$1017.84$

11. $\left(1 + \dfrac{.06}{12}\right)^{12}(\$2000) = \$2123.36$

$$\left[\dfrac{\left(1 + \frac{.06}{12}\right)^{12} - 1}{\frac{.06}{12}}\right](\$200) = \$2467.11$$

$200 each month is better, by $343.75.

13. $\left[\dfrac{\frac{.04}{2}}{\left(1 + \frac{.04}{2}\right)^{2 \times 15} - 1}\right](\$1,000,000) = \$24,649.92$

15. Jack withdraws for $12(.75) = 9$ months.

$$\left[\dfrac{\left(1 + \frac{.06}{12}\right)^{12 \times .75} - 1}{\frac{.06}{12}\left(1 + \frac{.06}{12}\right)^{12 \times .75}}\right](\$100) = \$877.91$$

17. $\left[\dfrac{\left(1 + \frac{.06}{12}\right)^{12 \times 10} - 1}{\frac{.06}{12}}\right](\$1000) = \$163,879.35$

$1000 at the end of each month is better.

19. $\left(1 + \dfrac{.08}{4}\right)^{4 \times 9}(\$1000) + \left[\dfrac{\left(1 + \frac{.08}{4}\right)^{4 \times 9} - 1}{\frac{.08}{4}}\right](\$100) = \$7239.32$

21. $\left[\dfrac{\left(1 + \frac{.06}{12}\right)^{12 \times 10} - 1}{\frac{.06}{12}}\right](\$100) + \left(1 + \dfrac{.06}{12}\right)^{12 \times 3}(\$1000) = \$17,584.62$

23. (a)

25. $.05P = \$1200$, so $P = \$24,000$.

27. $P = (1.06)^7(\$10,000) = \$15,036.30$

$$R = \left[\dfrac{.06(1.06)^4}{(1.06)^4 - 1}\right](\$15,036.30) = \$4339.35$$

29. $P = \dfrac{R}{i} = \dfrac{\$6000}{.06} = \$100,000$

$$\dfrac{\$100,000}{(1.06)^9} = \$59,189.85$$

31. a. $1

b. The second account is an annuity with $R = i(\$1) = i$ and $F = s_{\overline{n}|i}i$

c. $s_{\overline{n}|i}i + 1 = (1+i)^n$; $s_{\overline{n}|i} = \dfrac{(1+i)^n - 1}{i}$

33. $s_{\overline{n+1}|i} = \dfrac{(1+i)^{n+1} - 1}{i} = \dfrac{(1+i)^{n+1} - (1+i) + i}{i} = \dfrac{(1+i)^{n+1} - (1+i)}{i} + \dfrac{i}{i} = (1+i)\dfrac{(1+i)^n - 1}{i} + 1$

$= (1+i)s_{\overline{n}|i} + 1$

35. Present value $= \left[\dfrac{(1.06)^{30} - 1}{.06(1.06)^{30}}\right]\left(\dfrac{.08}{2}\right)(\$5000) + \dfrac{\$5000}{(1.06)^{30}} = \$3623.52$

37. a. $\left[\dfrac{\frac{.18}{12}\left(1 + \frac{.18}{12}\right)^{12 \times 5}}{\left(1 + \frac{.18}{12}\right)^{12 \times 5} - 1}\right](\$50,000) = \$1269.67$

b. Treat as two annuities. The first has a present value of

$\left[\dfrac{(1.01)^{60} - 1}{.01(1.01)^{60}}\right](.015)(\$50,000) = \$33,716.28.$

The second has a present value of

$\left[\dfrac{(1.01)^{60} - 1}{.01(1.01)^{60}}\right](\$1269.67) \div (1.01)^{60} = \$31,418.60.$

The total present value = \$65,134.88

39. Set up a table with $Y_1 = (1.05^{\wedge} X - 1)/.05 * 1000$. After 1, 2, and 3 years, Y_1 equals \$1000, \$2050, and \$3152.50. Y_1 equals \$30,539 after 19 years, and exceeds \$50,000 after 26 years.

41. Use $Y_1 = (1.001^{\wedge} X - 1)/.001 * 15$. $Y_1 = 503$ after 33 weeks.

Exercises 3

1. $R = \left[\dfrac{\frac{.06}{12}\left(1 + \frac{.06}{12}\right)^{12 \times 5}}{\left(1 + \frac{.06}{12}\right)^{12 \times 5} - 1}\right](\$10,000) = \$193.33$

3. $P = \left[\dfrac{\left(1 + \frac{.12}{2}\right)^{2 \times 10} - 1}{\frac{.12}{2}\left(1 + \frac{.12}{2}\right)^{2 \times 10}}\right](\$1000) = \$11,469.92$

5. a. $\dfrac{.12}{12}(\$58,331) = \$583.31$

b. $\$600 - \$583.31 = \$16.69$

   c. $58,331 − $16.69 = $58,314.31

   d. $\left[\dfrac{\left(1+\frac{.12}{12}\right)^{12(30-25)}-1}{\frac{.12}{12}\left(1+\frac{.12}{12}\right)^{12(30-25)}}\right]($600) = $26,973.02$

   e. $\left[\dfrac{\left(1+\frac{.12}{12}\right)^{12(30-26)}-1}{\frac{.12}{12}\left(1+\frac{.12}{12}\right)^{12(30-26)}}\right]($600) = $22,784.38$

      $26,973.02 − $22,784.38 = $4188.65
      (With a $.01 discrepancy due to rounding errors)

   f. Use result of part (d): .01($26,973.02) = $269.73

7. a. $\left[\dfrac{\frac{.12}{12}\left(1+\frac{.12}{12}\right)^{12\times3}}{\left(1+\frac{.12}{12}\right)^{12\times3}-1}\right]($8000) = 265.71

   b. ($265.71)(3 × 12) = $9565.56

   c. $9565.56 − $8000 = $1565.56

   d. $\left[\dfrac{\left(1+\frac{.12}{12}\right)^{12(3-1)}-1}{\frac{.12}{12}\left(1+\frac{.12}{12}\right)^{12(3-1)}}\right]($265.71) = 5644.58

   e. $\left[\dfrac{\left(1+\frac{.12}{12}\right)^{12(3-2)}-1}{\frac{.12}{12}\left(1+\frac{.12}{12}\right)^{12(3-2)}}\right]($265.71) = 2990.59

   f. 12($265.71) − ($5644.58 − $2990.59) = $534.53

   g.

Payment number	Amount	Interest	Applied to Principal	Unpaid balance
1	$265.71	$80.00	$185.71	$7814.29
2	265.71	78.14	187.57	7626.72
3	265.71	76.27	189.44	7437.28
4	265.71	74.37	191.34	7245.94

9. $\left(1+\dfrac{.09}{12}\right)($10,000) − $1125 = 8950

11. $\left[\dfrac{\left(1+\frac{.12}{2}\right)^{2\times8}-1}{\frac{.12}{2}\left(1+\frac{.12}{2}\right)^{2\times8}}\right](\$1000)+\left[\dfrac{1}{\left(1+\frac{.12}{2}\right)^{2\times8}}\right](\$10,000)=\$14,042.36$

13. $\left[\dfrac{\frac{.12}{12}\left(1+\frac{.12}{12}\right)^{4}}{\left(1+\frac{.12}{12}\right)^{4}-1}\right]100=256.28$

Payment	Amount	Interest	Applied to Principal	Unpaid balance
1	$256.28	$10.00	$246.28	$753.72
2	256.28	7.54	248.74	504.98
3	256.28	5.05	251.23	253.74
4	256.28	2.54	253.74	0.00

15. $\left[\dfrac{\frac{.09}{12}\left(1+\frac{.09}{12}\right)^{12\times30}}{\left(1+\frac{.09}{12}\right)^{12\times30}-1}\right](\$120,000-\$20,000)=\$804.62$

17. $\left[\dfrac{\left(1+\frac{.09}{12}\right)^{12\times25}-1}{\frac{.09}{12}\left(1+\frac{.09}{12}\right)^{12\times25}}\right](\$1200-\$200)=\$119,161.62$

19. $\left[\dfrac{\left(1+\frac{.06}{12}\right)^{12\times3}-1}{\frac{.06}{12}\left(1+\frac{.06}{12}\right)^{12\times3}}\right](\$100)=\$3287.10$

$\$6287.10-(\$2000+\$3287.10)=\$1000.00$

$\left(1+\dfrac{.06}{12}\right)^{12\times3}(\$1000)=\$1196.68$

21. (a)

23. **a.** $\left[\dfrac{(1.06)^{20}-1}{.06}\right](\$5000)=\$183,927.96$

     b. $\left[\dfrac{.06(1.06)^{10}}{(1.06)^{10}-1}\right](\$183,927.96)=\$24,989.92$

     c. $\left[\dfrac{(1.06)^{10-5}-1}{.06(1.06)^{10-5}}\right](\$24,989.92)=\$105,266.63$

25. $\left[\dfrac{\left(1+\frac{.12}{12}\right)^{12\times15}-1}{\frac{.12}{12}\left(1+\frac{.12}{12}\right)^{12\times15}}\right]$($30 million) = \$2.5 billion

27. Future value needs to be $(1.06)^{10}$($6 million) = \$10.745 million.

$\left[\dfrac{\left(1+\frac{.12}{12}\right)^{12\times10}-1}{\frac{.12}{12}}\right]$($100,000) = \$23.004 million

The sinking fund will be adequate.

29. After 1 month:

$\left(1+\dfrac{.09}{12}\right)$($2188.91) - \$100 = \$2105.33

After 2 months:

$\left(1+\dfrac{.09}{12}\right)$($2105.33) - \$100 = \$2021.12

After 3 months:

$\left(1+\dfrac{.09}{12}\right)$($2021.12) - \$100 = \$1936.28

The loan will be paid off after 24 months.

31. Enter 10000, then run

1.0075 * Ans − 166.68 repeatedly. After 40 iterations, the balance is \$5741.79, which means \$4258.21 has been paid off. The balance drops below \$5000 after 46 months.

33. The balance B must drop to where $\left(\dfrac{.085}{12}\right)B \le \$250$, or $B \le \$35{,}294.12$. Let $Y_1 = Y_6(300-X)*1000$

where $Y_6 = ((1+I)\wedge X - 1)/(I(1+I)\wedge X)$. Make a table. Then $B \le \$35{,}294.12$ after 260 months.

Chapter 10 Supplementary Exercises

1. (b)

2. $\left[\dfrac{\frac{.06}{12}}{\left(1+\frac{.06}{12}\right)^{12\times10}-1}\right]$($80,000) = \$488.16

3. The monthly mortgage payment should not exceed $\left(\dfrac{19,200}{12}\right)(.25) = \$400$.

$\left[\dfrac{\left(1+\frac{.09}{12}\right)^{12\times30}-1}{\frac{.09}{12}\left(1+\frac{.09}{12}\right)^{12\times30}}\right]$($400) = \$49,712.75

4. $50\left(1+\dfrac{.073}{365}\right)^{365} = \$53.79$

5. 9% compounded daily yields $\left(1+\dfrac{.09}{365}\right)^{365} - 1 = .0942 = 9.42\%$ annually. 10% compounded annually is better.

6. $\left[\dfrac{\left(1+\frac{.06}{12}\right)^{12\times5} - 1}{\frac{.06}{12}}\right](\$200) = \$13,954.01$

7. a. $\left[\dfrac{\frac{12}{12}\left(1+\frac{12}{12}\right)^{12\times15}}{\left(1+\frac{12}{12}\right)^{12\times15} - 1}\right](\$200,000) = \$2400.34$

b. $\left[\dfrac{\left(1+\frac{12}{12}\right)^{12(15-5)} - 1}{\frac{12}{12}\left(1+\frac{12}{12}\right)^{12(15-5)}}\right](\$2400.34) = \$167,304.68$

8. $(\$24,000)(1.005)^{120} = \$43,665.52$

9. $\dfrac{\$50,000}{\left(1+\frac{.06}{12}\right)^{12\times10}} = \$27,481.64$

10. $\dfrac{\$10,000}{\left(1+\frac{.06}{12}\right)^{12\times2}} + \dfrac{\$5000}{\left(1+\frac{.06}{12}\right)^{12\times3}} = \$13,050.08$

11. $\left[\dfrac{\frac{.06}{12}\left(1+\frac{.06}{12}\right)^{12\times4}}{\left(1+\frac{.06}{12}\right)^{12\times4} - 1}\right](\$12,000 - \$3,000) = \$211.37$

12. $\left[\dfrac{\frac{.04}{2}\left(1+\frac{.04}{2}\right)^{2\times5}}{\left(1+\frac{.04}{2}\right)^{2\times5} - 1}\right]\left(1+\dfrac{.04}{2}\right)^{2\times2}(\$100,000) = \$12,050.34$

13. $\dfrac{\$30,000}{\left(1+\frac{.06}{12}\right)^{12\times15}} = \$12,224.47$

$\$105,003.50 - \$12,224.47 = \$92,779.03$

$\left[\dfrac{\frac{.06}{12}\left(1+\frac{.06}{12}\right)^{12\times15}}{\left(1+\frac{.06}{12}\right)^{12\times15} - 1}\right](\$92,779.03) = \$782.92$

14. $\dfrac{\$100,000}{\left(1+\frac{.12}{12}\right)^{12\times10}} = \$30,299.48$

$\left[\dfrac{\frac{.12}{12}\left(1+\frac{.12}{12}\right)^{12\times10}}{\left(1+\frac{.12}{12}\right)^{12\times10}-1}\right](\$509,289.22-\$30,299.48) = \$6872.11$

15. $\left[\dfrac{\left(1+\frac{.06}{12}\right)^{12\times30}-1}{\frac{.06}{12}}\right](\$100) = \$100,451.50$

16. $\left[\dfrac{\left(1+\frac{.12}{12}\right)^{12\times10}-1}{\frac{.12}{12}\left(1+\frac{.12}{12}\right)^{12\times10}}\right](\$2000) = \$139,401.04$

17. Investment A: $\left[\dfrac{(1+.06)^{10}-1}{.06}\right]1000 = \$13,180.79$

Investment B: $5000(1+.06)^{5}+5000 = \$11,691.13$
Thus Investment A is the better investment.

18. Present value of annuity is $\left[\dfrac{\left(1+\frac{.09}{12}\right)^{12\times5}-1}{\frac{.09}{12}\left(1+\frac{.09}{12}\right)^{12\times5}}\right](\$5) = \$240.87.$

The present value of $1000 is $\dfrac{\$1000}{\left(1+\frac{.09}{12}\right)^{12\times5}} = \$638.70.$

Yes, it is a bargain, since the present value is $240.87 + 638.70 = \$879.57$.

19. $\left(1+\dfrac{.10}{2}\right)^{2}-1 = .1025 = 10.25\%$

20. $\left(1+\dfrac{.18}{12}\right)^{12}-1 = .1956 = 19.56\%$

21. $\left(1+\dfrac{.08}{4}\right)^{4\times15}(\$10,000)+\left[\dfrac{\left(1+\frac{.08}{4}\right)^{4\times15}-1}{\frac{.08}{4}}\right](\$1000) = \$146,861.85$

22. $R = \left[\dfrac{\frac{.06}{12}\left(1+\frac{.06}{12}\right)^{36}}{\left(1+\frac{.06}{12}\right)^{36}-1}\right]10,000 = 304.22$

Payment number	Amount	Interest	Applied to Principal	Unpaid balance
1	$304.22	$50.00	$254.22	$9745.78
2	304.22	48.73	255.49	9490.29
3	304.22	47.45	256.77	9233.52
4	304.22	46.17	258.05	8975.47
5	304.22	44.88	259.34	8716.13
6	304.22	43.58	260.64	8455.49

23. $\left[\dfrac{(1.01)^{120} - 1}{.01} \right] (\$200)(1.01)^{120} = \$151,843.34$

24. $\left[\dfrac{\frac{.06}{12}\left(1 + \frac{.06}{12}\right)^{12 \times 5}}{\left(1 + \frac{.06}{12}\right)^{12 \times 5} - 1} \right] (\$300,000) = \$5799.84$

25. $\left[\dfrac{\frac{.09}{12}\left(1 + \frac{.09}{12}\right)^{12 \times 30}}{\left(1 + \frac{.09}{12}\right)^{12 \times 30} - 1} \right] (\$150,000) = \$1206.93$

Chapter 11

Exercises 1

1. $a = 4$, $b = -6$, $\dfrac{b}{1-a} = \dfrac{-6}{1-4} = 2$

3. $a = -\dfrac{1}{2}$, $b = 0$, $\dfrac{b}{1-a} = \dfrac{0}{1+\frac{1}{2}} = 0$

5. $a = -\dfrac{2}{3}$, $b = 15$, $\dfrac{b}{1-a} = \dfrac{15}{1+\frac{2}{3}} = 9$

7. a. $y_0 = 10$, $y_1 = \dfrac{1}{2}(10) - 1 = 4$,

$y_2 = \dfrac{1}{2}(4) - 1 = 1$, $y_3 = \dfrac{1}{2}(1) - 1 = -\dfrac{1}{2}$,

$y_4 = \dfrac{1}{2}\left(-\dfrac{1}{2}\right) - 1 = -\dfrac{5}{4}$

b.

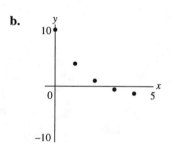

c. $y_n = \dfrac{-1}{1-\frac{1}{2}} + \left(10 - \dfrac{-1}{1-\frac{1}{2}}\right)\left(\dfrac{1}{2}\right)^n$

$= -2 + 12\left(\dfrac{1}{2}\right)^n$

9. a. $y_0 = 3.5$, $y_1 = 2(3.5) - 3 = 4$,

$y_2 = 2(4) - 3 = 5$, $y_3 = 2(5) - 3 = 7$,

$y_4 = 2(7) - 3 = 11$

b.

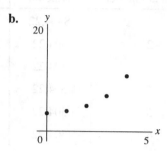

c. $y_n = \dfrac{-3}{1-2} + \left(3.5 - \dfrac{-3}{1-2}\right)(2)^n$

$= 3 + (.5)2^n$

11. a. $y_0 = 17.5$, $y_1 = -.4(17.5) + 7 = 0$,

$y_2 = -.4(0) + 7 = 7$,

$y_3 = -.4(7) + 7 = 4.2$,

$y_4 = -.4(4.2) + 7 = 5.32$

b.

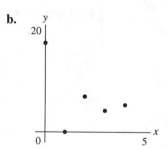

c. $y_n = \dfrac{7}{1+.4} + \left(17.5 - \dfrac{7}{1+.4}\right)(-.4)^n$

$= 5 + 12.5(-.4)^n$

13. a. $y_0 = 15$, $y_1 = 2(15) - 16 = 14$,

$y_2 = 2(14) - 16 = 12$,

$y_3 = 2(12) - 16 = 8$,

$y_4 = 2(8) - 16 = 0$

b.

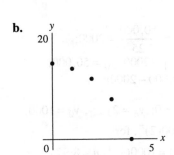

c. $y_n = \dfrac{-16}{1-2} + \left(15 - \dfrac{-16}{1-2}\right)(2)^n$

$= 16 - 2^n$

15. $y_0 = 6 - 5(.2)^0 = 1$

$y_1 = 6 - 5(.2)^1 = 5$

$y_2 = 6 - 5(.2)^2 = 5.8$

$y_3 = 6 - 5(.2)^3 = 5.96$

$y_4 = 6 - 5(.2)^4 = 5.992$

17. $y_n = 1.05 y_{n-1}, \; y_0 = 1000$

19. $y_n = 1.05 y_{n-1} + 100, \; y_0 = 1000$

21. a. $y_0 = 1, \; y_1 = 1 + 2 = 3, \; y_2 = 3 + 2 = 5,$
$y_3 = 5 + 2 = 7, \; y_4 = 7 + 2 = 9$

b.

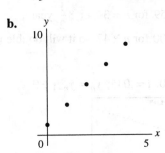

c. $a = 1$, so the denominator of $\dfrac{b}{1-a}$ is zero.

23. $1.20(55) - 36 = \$30$

25.

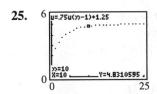

$y_{10} = 4.8310595; \; 5 - y_n < .01$ for $n \ge 20$

27.

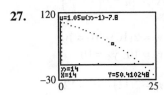

$y_{14} = 50.41; \; y_n = 0$ for $n = 22$

29.

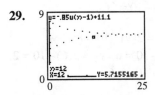

$y_{12} = 5.7155165; \; |6 - y_n| \le .1$ for $n \ge 19$.

31.

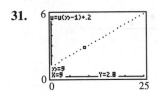

$y_9 = 2.8; \; y_n = 4.6$ for $n = 18$

Exercises 2

1. $b = 5, \; y_0 = 1$
From formula (2), $y_n = 1 + 5n$.

3. $y_0 = 80, \; i = \dfrac{.09}{12} = .0075, \; n = 5 \times 12 = 60$
From formula (4), $y_n = 80(1.0075)^{60}$

5. $y_0 = 80, \; i = \dfrac{1}{365}, \; n = 5 \times 365 = 1825$
From formula (4), $y_n = 80\left(1 + \dfrac{1}{365}\right)^{1825}$.

7. $y_0 = 80$, $i = .07$, $n = 5$
From formula (3),
$y_n = 80 + .07 \times 80 \times 5 = 108$

9. $y_0 = A$, $i = \dfrac{r}{k}$, $n = kt$

From formula (4), $y_n = A\left(1 + \dfrac{r}{k}\right)^{kt}$.

11. a. $y_0 = 10$; $y_1 = 2(10) - 10 = 10$;
$y_2 = 10$; $y_3 = 10$; $y_4 = 10$

b. $y_0 = 11$; $y_1 = 2(11) - 10 = 12$;
$y_2 = 2(12) - 10 = 14$;
$y_3 = 2(14) - 10 = 18$;
$y_4 = 2(18) - 10 = 26$

c. $y_0 = 9$; $y_1 = 2(9) - 10 = 8$;
$y_2 = 2(8) - 10 = 6$; $y_3 = 2(6) - 10 = 2$;
$y_4 = 2(2) - 10 = -6$

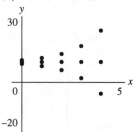

13. $a = .4$, $b = 3$;
$$y_n = \dfrac{3}{1 - .4} + \left(7 - \dfrac{3}{1 - .4}\right)(.4)^n$$
$$= 5 + 2(.4)^n;$$
as n gets large, y_n approaches 5.

15. $a = -5$, $b = 0$;
$$y_n = \dfrac{0}{1 + 5} + \left(2 - \dfrac{0}{1 + 5}\right)(-5)^n = 2(-5)^n;$$
as n gets large, y_n gets arbitrarily large, alternating between being positive and negative.

17. $a = 1 + \dfrac{.09}{12} = 1.0075$, $b = -350$;
$y_n = 1.0075 y_{n-1} - 350$, $y_0 = 38,900$

19. $a = 1$, $b = \dfrac{50,000}{25} = 2000$;
$y_n = y_{n-1} - 2000$, $y_0 = 50,000$;
$y_n = 50,000 - 2000n$

21. $a = 2$, $b = 0$; $y_n = 2y_{n-1}$, $y_0 = 1000$;
$y_n = 1000(2)^n$; for
$500,000 = 1000(2)^n$, $n \approx 8.97$.
8.97×2 min ≈ 18 min
(E)

23. $a = 2$, $b = 0$; $y_n = 2y_{n-1}$; $y_n = y_0(2)^n$; for
$y_6 = 640 = y_0(2)^6$, $y_0 = \dfrac{640}{2^6} = 10$.
Then $y_3 = 10(2)^3 = \$80$.
(A)

25. $y_0 = 1000$, $i = \dfrac{.06}{4} = .015$; $y_n = 1.015 y_{n-1}$

For $n = 3 \times 4 = 12$, $y_{12} = \$1195.62$;

$y_n = 1659$ for $n = 34$, or $8\dfrac{1}{2}$ years;

$y_n \geq 2000$ for $n \geq 47$, so it will double in 47 quarters.

27. $y_0 = 200$, $i = .045$; $y_n = y_{n-1} + 9$

For $n = 5$, $y_5 = \$245$;
$y_n \geq 308$ for $n \geq 12$ years;
$y_n \geq 400$ for $n \geq 23$ years

29.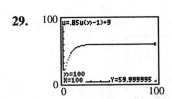

y_n approaches 60.

Exercises 3

1. (a), (b), (d), (f), (h)

3. (b), (d), (e), (f)

5. (b), (d), (e), (f)

7. (a), (c), (h), and possibly (g)

9. Possible answer:

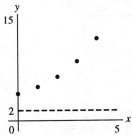

11. Possible answer:

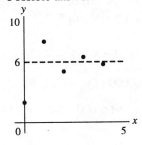

13. Possible answer:

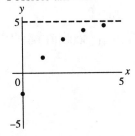

15. Draw $y = \dfrac{b}{1-a} = -2$ as a dashed line.

$a = 3 > 0$, so the graph is monotonic.

$|a| = 3 > 1$, so the graph is repelled from $y = -2$.

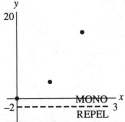

17. Draw $y = \dfrac{b}{1-a} = 6$ as a dashed line.

$y_0 = 6$, so the graph is constant.

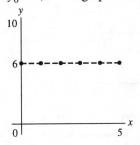

19. Draw $y = \dfrac{b}{1-a} = 4$ as a dashed line.

$a = -2 < 0$, so the graph is oscillating.

$|a| = 2 > 1$, so the graph is repelled from $y = 4$.

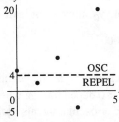

21. Draw $y = \dfrac{b}{1-a} = 10,000$ as a dashed line.

$a = .7 > 0$, so the graph is monotonic.

$|a| = .7 < 1$, so the graph is attracted to

$y = 10,000$.

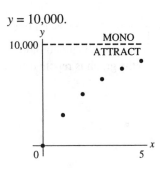

23. Draw $y = \dfrac{b}{1-a} = 1$ as a dashed line.

$a = -.6 < 0$, so the graph is oscillating.

$|a| = .6 < 1$, so the graph is attracted to $y = 1$.

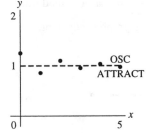

25. $a = 1 + i = 1.0075$, $b = -450$

The loan, y_0, must be less than

$$\dfrac{b}{1-a} = \$60,000.$$

27. $a = 1 + i = 1.06$, $b = -120$

a. $y_n = 1.06 y_{n-1} - 120$

b. The deposit, y_0, must be at least

$$\dfrac{b}{1-a} = \$2000.$$

For Exercises 29–34, choose y_0 as the beginning value; choose $a > 0$ or < 0 depending on whether the graph is monotonic or oscillating, also $|a| > 1$ or < 1 depending on whether the graph is unbounded or approaches a value; and if it approaches or is repelled from a value, choose b so that $\dfrac{b}{1-a}$ equals that value.

29. Possible answer: $y_n = .5 y_{n-1} + 4$, $y_0 = 1$

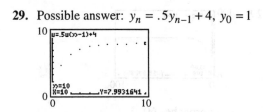

31. Possible answer: $y_n = 2 y_{n-1}$, $y_0 = 1$

33. Possible answer: $y_n = -2 y_{n-1}$, $y_0 = 5$

Exercises 4

1. $a = 1 + i = 1.0075$, $b = -261.50$;
$y_n = 1.0075 y_{n-1} - 261.50$, $y_0 = 32,500$

3. $a = 1 + i = 1.015$, $b = 200$;
$y_n = 1.015 y_{n-1} + 200$, $y_0 = 4000$

5. $a = 1 + i = 1.01$, $b = -660$, $\dfrac{b}{1-a} = 66,000$

$y_{120} = 0 = 66,000 + (y_0 - 66,000)(1.01)^{120}$

$y_0 = \dfrac{-66,000}{(1.01)^{120}} + 66,000 \approx \$46,002.34$

7. $a = 1 + i = 1.06$, $b = 300$,

$\dfrac{b}{1-a} = -5000$, $y_0 = 0$

$y_{20} = -5000 + [0 - (-5000)](1.06)^{20}$
$\approx \$11,035.68$

9. $a = 1 + i = 1.02, b = 0$

$$y_{56} = 6000 = y_0(1.02)^{56}$$

$$y_0 = \frac{6000}{(1.02)^{56}} \approx \$1979.44$$

11. $a = 1 + i = 1.01, y_0 = 4000$

$$y_{36} = 0 = \frac{b}{-.01} + \left(4000 - \frac{b}{-.01}\right)(1.01)^{36}$$

$$= 4000(1.01)^{36} + 100[(1.01)^{36} - 1]b$$

$$b = \frac{-4000(1.01)^{36}}{100[(1.01)^{36} - 1]} \approx -132.86$$

$$\$132.86$$

13. $a = 1 + i = 1.08, b = -4$
Use nMin $= 0$, $u(n) = 1.08u(n-1)-4$,
$u(n$Min$)=\{45\}$
$u(10) \approx \$39.205$ million
$u(n) \le 0$ for $n = 30$ years

15. $a = 1 + i = 1.005, b = 100$
Use nMin $= 0$, $u(n) = 1.005u(n-1) + 100$,
$u(n$Min$) = \{0\}$
$u(5) \approx \$505.03$
$u(10) \approx \$1022.80$
$u(15) \approx \$1553.65$
$u(n) \ge \$3228$ for $n = 30$
$u(n) > \$4000$ for $n = 37$

Exercises 5

1. $a = 1 + .03 - .01 = 1.02, b = 0, \dfrac{b}{1-a} = 0$

$y_n = 1.02y_{n-1}, y_0 = 100$ million
$a > 0$: monotonic; $|a| > 1$: repelled from $y = 0$

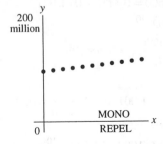

3. $a = 1 - .25 = .75, b = 0, \dfrac{b}{1-a} = 0$

$y_n = .75y_{n-1}$
$a > 0$: monotonic; $|a| < 1$: attracted to $y = 0$

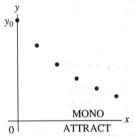

5. $y_n = y_{n-1} + .08(100 - y_{n-1}) = .92y_{n-1} + 8,$
$y_0 = 0$

$a = .92, b = 8, \dfrac{b}{1-a} = 100$
$a > 0$: monotonic; $|a| < 1$: attracted to
$y = 100$

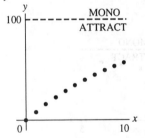

7. $y_n = y_{n-1} + .30(12 - y_{n-1})$
$= .7y_{n-1} + 3.6, y_0 = 0$

$a = .7, b = 3.6, \dfrac{b}{1-a} = 12$
$a > 0$: monotonic; $|a| < 1$: attracted to $y = 12$

9. $a = 1 + i = 1.05, b = -1000, \dfrac{b}{1-a} = 20,000$

$y_n = 1.05 y_{n-1} - 1000, \ y_0 = 30,000$

$a > 0$: monotonic; $|a| > 1$: repelled from $y = 20,000$

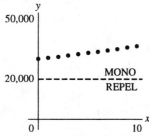

11. $y_n = y_{n-1} + .20(70 - y_{n-1}) = .8 y_{n-1} + 14$, $y_0 = 40$

$a = .8, b = 14, \dfrac{b}{1-a} = 70$

$a > 0$: monotonic; $|a| < 1$: attracted to $y = 70$

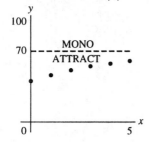

13. $p_n = 20 - .1 q_n = 20 - .1(5 p_{n-1} - 10)$

$= -.5 p_{n-1} + 21, \ p_0 = 10$

$a = -.5, b = 21, \dfrac{b}{1-a} = 14$

$a < 0$: oscillating, $|a| < 1$: attracted to $y = 14$

15. Use nMin$=0$,

$u(n) = (1 + .035 - .02)u(n-1) - .0003$,

$u(n$Min$) = \{5\}$

$u(5) \approx 5.38$ million

$u(n) > 6$ for $n = 13$, or in the year 2003

$u(n) \geq 10$ for $n = 47$, or in the year 2037

Chapter 11 Supplementary Exercises

1. a. $y_1 = -3(1) + 8 = 5; \ y_2 = -3(5) + 8 = -7;$
$y_3 = -3(-7) + 8 = 29$

b. $a = -3, b = 8$

$y_n = \dfrac{b}{1-a} + \left(y_0 - \dfrac{b}{1-a} \right) a^n$

$= 2 + (1 - 2)(-3)^n = 2 - (-3)^n$

c. $2 - (-3)^4 = -79$

2. a. $y_1 = 10 - \dfrac{3}{2} = \dfrac{17}{2}; \ y_2 = \dfrac{17}{2} - \dfrac{3}{2} = 7;$
$y_3 = 7 - \dfrac{3}{2} = \dfrac{11}{2}$

b. $a = 1, \ b = -\dfrac{3}{2}$

$y_n = 10 - \dfrac{3}{2} n$

c. $10 - \dfrac{3}{2}(6) = 1$

3. $a = 1 + i = 1.02, b = 0, \dfrac{b}{1-a} = 0$

$y_{40} = 6600 = 0 + (y_0 - 0)(1.02)^{40}$

$= y_0 (1.02)^{40}$

$y_0 = \dfrac{6600}{(1.02)^{40}} \approx \$2989.08$

4. $a = 1 + i = 1.001, b = 0, \dfrac{b}{1-a} = 0$,

$y_0 = 1000$

$y_{104} = 0 + (1000 - 0)(1.001)^{104}$

$= \$1109.54$

5. Draw $y = \dfrac{b}{1-a} = 6$ as a dashed line.

$a = -\dfrac{1}{3} < 0$, so the graph is oscillating.

$|a| = \dfrac{1}{3} < 1$, so the graph is attracted to $y = 6$.

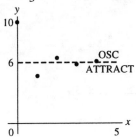

6. Draw $y = \dfrac{b}{1-a} = 4$ as a dashed line.

$a = 1.5 > 0$, so the graph is monotonic.

$|a| = 1.5 > 1$, so the graph is repelled from $y = 4$.

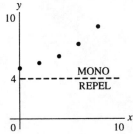

7. a. $y_n = 1.03 y_{n-1} - 600,\ y_0 = 120,000$

   b. $\dfrac{b}{1-a} = 20,000;\ y_{20} = 20,000 + (120,000 - 20,000)(1.03)^{20} \approx 200,611$

8. a. $y_n = 1.01 y_{n-1} - 360,\ y_0 = 35,000$

   b. $\dfrac{b}{1-a} = 36,000;\ y_{84} = 36,000 + (35,000 - 36,000)(1.01)^{84} \approx \$33,693.28$

9. $a = 1 + i = 1.001,\ y_0 = 0,\ 21 \times 52 = 1092$

$$y_{1092} = 40,000 = \frac{b}{-.001} + \left(0 - \frac{b}{-.001}\right)(1.001)^{1092} = 1000[(1.001)^{1092} - 1]b$$

$$b = \frac{40,000}{1000[(1.001)^{1092} - 1]} \approx \$20.22$$

10. $a = 1 + i = 1.005$, $y_0 = 33,100$

$y_{240} = 0$

$= \dfrac{b}{-.005} + \left(33,100 - \dfrac{b}{-.005}\right)(1.005)^{240}$

$= 33,100(1.005)^{240} + 200[(1.005)^{240} - 1]b$

$b = \dfrac{-33,100(1.005)^{240}}{200[(1.005)^{240} - 1]} \approx -237.14$

$237.14

11. $a = 1 + i = 1.08$, $b = -2400$,

$\dfrac{b}{1-a} = \$30,000$

$y_{18} = 0 = 30,000 + (y_0 - 30,000)(1.08)^{18}$

$y_0 = \dfrac{-30,000}{(1.08)^{18}} + 30,000 \approx \$22,492.53$

12. $y_n = 1.005 y_{n-1} + 5$, $y_0 = 0$

$a = 1.005$, $b = 5$, $\dfrac{b}{1-a} = -1000$

$y_{48} = -1000 + [0 - (-1000)](1.005)^{48}$

$\approx \$270.49$

13. $y_n = y_{n-1} + .10(1,000,000 - y_{n-1})$

$= .9 y_{n-1} + 100,000$, $y_0 = 0$

$a = .9$, $b = 100,000$, $\dfrac{b}{1-a} = 1,000,000$

$a > 0$: monotonic, $|a| < 1$: attracted to

$y = 1,000,000$

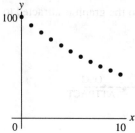

14. $y_n = y_{n-1} - .08 y_{n-1} = .92 y_{n-1}$, $y_0 = 100$

$a = .92$, $b = 0$, $\dfrac{b}{1-a} = 0$

$a > 0$: monotonic; $|a| < 1$: attracted to $y = 0$

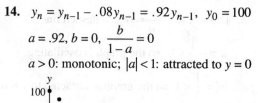

Chapter 12

Exercises 1

1. Statement

3. Statement

5. Not a statement—not a declarative sentence.

7. Not a statement—not a declarative sentence.

9. Statement

11. Not a statement—x is not specified.

13. Not a statement—not a declarative sentence.

15. Statement

17. **a.** Ozone is opaque to ultraviolet light, and life on earth requires ozone.

    b. Ozone is not opaque to ultraviolet light, or life on earth requires ozone.

    c. Ozone is not opaque to ultraviolet light, or life on earth does not require ozone.

    d. It is not the case that life on earth does not require ozone.

19. **a.** Florida borders Alabama or Florida borders Mississippi: $p \vee q$

    b. $p \wedge \sim q$

    c. $q \wedge \sim p$

    d. $\sim p \wedge \sim q$

Exercises 2

1.

p	q		p	$\wedge$	$\sim$	q
T	T		T	**F**	F	T
T	F		T	**T**	T	F
F	T		F	**F**	F	T
F	F		F	**F**	T	F
(1)	(2)			(4)	(3)	

3.

p	q	r	(p	∧	~	r)	∨	q
T	T	T	T	F	F	T	**T**	T
T	T	F	T	T	T	F	**T**	T
T	F	T	T	F	F	T	**F**	F
T	F	F	T	T	T	F	**T**	F
F	T	T	F	F	F	T	**T**	T
F	T	F	F	F	T	F	**T**	T
F	F	T	F	F	F	T	**F**	F
F	F	F	F	F	T	F	**F**	F
(1)	(2)	(3)		(5)	(4)		(6)	

5.

p	q	r	~	[(p	∧	r)	∨	q]
T	T	T	**F**	T	T	T	T	T
T	T	F	**F**	T	F	F	T	T
T	F	T	**F**	T	T	T	T	F
T	F	F	**T**	T	F	F	F	F
F	T	T	**F**	F	F	T	T	T
F	T	F	**F**	F	F	F	T	T
F	F	T	**T**	F	F	T	F	F
F	F	F	**T**	F	F	F	F	F
(1)	(2)	(3)	(6)		(4)		(5)	

7.

p		p	∧	~	p
T		T	**T**	F	T
F		F	**T**	T	F
(1)			(3)	(2)	

9.

p	q	r		p	⊕	(q	∧	r)
T	T	T		T	**F**	T	T	T
T	T	F		T	**F**	T	T	F
T	F	T		T	**F**	F	T	T
T	F	F		T	**T**	F	F	F
F	T	T		F	**T**	T	T	T
F	T	F		F	**T**	T	T	F
F	F	T		F	**T**	F	T	T
F	F	F		F	**F**	F	F	F
(1)	(2)	(3)			(5)		(4)	

11.

p	q	r		(p	∨	q)	∧	(p	∨	r)
T	T	T		T	T	T	**T**	T	T	T
T	T	F		T	T	T	**T**	T	T	F
T	F	T		T	T	F	**T**	T	T	T
T	F	F		T	T	F	**T**	T	T	F
F	T	T		F	T	T	**T**	F	T	T
F	T	F		F	T	T	**F**	F	F	F
F	F	T		F	F	F	**F**	F	T	T
F	F	F		F	F	F	**F**	F	F	F
(1)	(2)	(3)			(4)		(6)		(5)	

13.

p	q		(p	∨	q)	∧	~	(p	∨	q)
T	T		T	T	T	**F**	F	T	T	T
T	F		T	T	F	**F**	F	T	T	F
F	T		F	T	T	**F**	F	F	T	T
F	F		F	F	F	**F**	T	F	F	F
(1)	(2)			(3)		(6)	(5)		(4)	

15.

p	q	r	~	(p	∨	q)	∧	r
T	T	T	F	T	T	T	**F**	T
T	T	F	F	T	T	T	**F**	F
T	F	T	F	T	T	F	**F**	T
T	F	F	F	T	T	F	**F**	F
F	T	T	F	F	T	T	**F**	T
F	T	F	F	F	T	T	**F**	F
F	F	T	T	F	F	F	**T**	T
F	F	F	T	F	F	F	**F**	F
(1)	(2)	(3)	(5)		(4)		(6)	

17.

p	q	r	~	p	∨	(q	∧	r)
T	T	T	F	T	**T**	T	T	T
T	T	F	F	T	**F**	T	F	F
T	F	T	F	T	**F**	F	F	T
T	F	F	F	T	**F**	F	F	F
F	T	T	T	F	**T**	T	T	T
F	T	F	T	F	**T**	T	F	F
F	F	T	T	F	**T**	F	F	T
F	F	F	T	F	**T**	F	F	F
(1)	(2)	(3)	(4)		(6)		(5)	

19.

p	q	~	p	∨	~	q	~	(p	∧	q)
T	T	F	T	**F**	F	T	**F**	T	T	T
T	F	F	T	**T**	T	F	**T**	T	F	F
F	T	T	F	**T**	F	T	**T**	F	F	T
F	F	T	F	**T**	T	F	**T**	F	F	F
(1)	(2)	(3)		(5)	(4)		(4)		(3)	

They are identical.

21.

p	*q*		*p*	⊕	*q*		(*p*	∨	*q*)	∧	~	(*p*	∧	*q*)
T	T		T	**F**	T		T	T	T	**F**	F	T	T	T
T	F		T	**T**	F		T	T	F	**T**	T	T	F	F
F	T		F	**T**	T		F	T	T	**T**	T	F	F	T
F	F		F	**F**	F		F	F	F	**F**	T	F	F	F
(1)	(2)			(3)				(4)		(6)	(5)		(3)	

They are identical.

23.

p	*q*	*r*		(*p*	∧	*q*)	∨	*r*		*p*	∧	(*q*	∨	*r*)
T	T	T		T	T	T	**T**	T		T	**T**	T	T	T
T	T	F		T	T	T	**T**	F		T	**T**	T	T	F
T	F	T		T	F	F	**T**	T		T	**T**	F	T	T
T	F	F		T	F	F	**F**	F		T	**F**	F	F	F
F	T	T		F	F	T	**T**	T		F	**F**	T	T	T
F	T	F		F	F	T	**F**	F		F	**F**	T	T	F
F	F	T		F	F	F	**T**	T		F	**F**	F	T	T
F	F	F		F	F	F	**F**	F		F	**F**	F	F	F
(1)	(2)	(3)			(4)		(5)				(5)		(4)	

(*p* ∧ *q*) ∨ *r* is T and *p* ∧ (*q* ∨ *r*) is F when *p* is F and *r* is T. Otherwise, the tables are identical.

25. a.

p		*p*	|	*p*
T		T	**F**	T
F		F	**T**	F
(1)			(2)	

b.

p	*q*		(*p*	|	*p*)	|	(*q*	|	*q*)
T	T		T	F	T	**T**	T	F	T
T	F		T	F	T	**T**	F	T	F
F	T		F	T	F	**T**	T	F	T
F	F		F	T	F	**F**	F	T	F
(1)	(2)			(3)		(5)		(4)	

c.

p	q	(p		q)			(p		q)
T	T	T	F	T	**T**	T	F	T
T	F	T	T	F	**F**	T	T	F
F	T	F	T	T	**F**	F	T	T
F	F	F	T	F	**F**	F	T	F
(1)	(2)		(3)		(5)		(4)	

d.

p	q	p		((p		q)		q)
T	T	T	**F**	T	F	T	T	T
T	F	T	**F**	T	T	F	T	F
F	T	F	**T**	F	T	T	F	T
F	F	F	**T**	F	T	F	T	F
(1)	(2)		(5)		(3)		(4)	

27. *p* has truth value T and *q* has truth value F.

a.

p	∨	~	q
T	**T**	T	F

b.

~	p	∧	q
F	T	**F**	F

c.

p	⊕	q
T	**T**	F

d.

~	p	⊕	q
F	T	**F**	F

e.

~	(p	⊕	q)
F	T	T	F

f.

(p	∨	q)	⊕	~	q
T	T	F	**F**	T	F

29. *p* has truth value F and *g* has truth value T.

a.

p	∧	~	q
F	**F**	F	T

b.

~	(p	⊕	q)
F	F	T	T

c.

q	∧	q
F	**F**	T

d.   ~   *p*   ∧   ~   *q*

      T   F   **F**   F   T

31. $(p \oplus q) \wedge r$

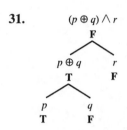

33. $(p \vee q) \wedge (p \vee \sim r)$

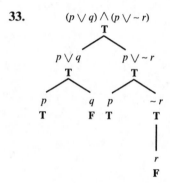

35. $(p$ ∧ $q)$ ∨   ~ p

     T   F   F   **F**   F   T

0 will be displayed.

37.

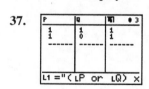

a.   T

b.   T

39.

a display showing:

P	q		◆ 3
1	1	1	
1	0	0	

L1 ="(ₗP and not(

a.   T

b.   F

41.

a display showing:

P	q		◆ 3
1	1	1	
1	0	0	
0	1	0	
0	0	1	

L1 ="not(ₗP xor ₗ

Exercises 3

1.

p	*q*		*p*	→	~	*q*
T	T		T	**F**	F	T
T	F		T	**T**	T	F
F	T		F	**T**	F	T
F	F		F	**T**	T	F
(1)	(2)			(4)	(3)	

3.

p	*q*		$(p$	$\oplus$	$q)$	→	*q*
T	T		T	F	T	**T**	T
T	F		T	T	F	**F**	F
F	T		F	T	T	**T**	T
F	F		F	F	F	**T**	F
(1)	(2)			(3)		(4)	

5.

p	q	r	(~	p	∧	q)	→	r
T	T	T	F	T	F	T	**T**	T
T	T	F	F	T	F	T	**T**	F
T	F	T	F	T	F	F	**T**	T
T	F	F	F	T	F	F	**T**	F
F	T	T	T	F	T	T	**T**	T
F	T	F	T	F	T	T	**F**	F
F	F	T	T	F	F	F	**T**	T
F	F	F	T	F	F	F	**T**	F
(1)	(2)	(3)	(4)		(5)		(6)	

7.

p	q	(p	→	q)	↔	(~	p	∨	q)
T	T	T	T	T	**T**	F	T	T	T
T	F	T	F	F	**T**	F	T	F	F
F	T	F	T	T	**T**	T	F	T	T
F	F	F	T	F	**T**	T	F	T	F
(1)	(2)		(4)		(6)	(3)		(5)	

9.

p	q	r	(p	→	q)	→	r
T	T	T	T	T	T	**T**	T
T	T	F	T	T	T	**F**	F
T	F	T	T	F	F	**T**	T
T	F	F	T	F	F	**T**	F
F	T	T	F	T	T	**T**	T
F	T	F	F	T	T	**F**	F
F	F	T	F	T	F	**T**	T
F	F	F	F	T	F	**F**	F
(1)	(2)	(3)	(4)			(5)	

11.

p	*q*	*r*	~	(*p*	∨	*q*)	→	(~	*p*	∧	*r*)
T	T	T	F	T	T	T	**T**	F	T	F	T
T	T	F	F	T	T	T	**T**	F	T	F	F
T	F	T	F	T	T	F	**T**	F	T	F	T
T	F	F	F	T	T	F	**T**	F	T	F	F
F	T	T	F	F	T	T	**T**	T	F	T	T
F	T	F	F	F	T	T	**T**	T	F	F	F
F	F	T	T	F	F	F	**T**	T	F	T	T
F	F	F	T	F	F	F	**F**	T	F	F	F
(1)	(2)	(3)	(6)		(4)		(8)	(5)		(7)	

13.

p	*q*	(*p*	∨	*q*)	↔	(*p*	∧	*q*)
T	T	T	T	T	**T**	T	T	T
T	F	T	T	F	**F**	T	F	F
F	T	F	T	T	**F**	F	F	T
F	F	F	F	F	**T**	F	F	F
(1)	(2)		(3)		(5)		(4)	

15.

p	*q*	*r*	[(*p*	∧	(*q*	∨	*r*)]	↔	[(*p*	∧	*q*)	∨	(*p*	∧	*r*)]
T	T	T	T	T	T	T	T	**T**	T	T	T	T	T	T	T
T	T	F	T	T	T	T	F	**T**	T	T	T	T	T	F	F
T	F	T	T	T	F	T	T	**T**	T	F	F	T	T	T	T
T	F	F	T	F	F	F	F	**T**	T	F	F	F	T	F	F
F	T	T	F	F	T	T	T	**T**	F	F	T	F	F	F	T
F	T	F	F	F	T	T	F	**T**	F	F	T	F	F	F	F
F	F	T	F	F	F	T	T	**T**	F	F	F	F	F	F	T
F	F	F	F	F	F	F	F	**T**	F	F	F	F	F	F	F
(1)	(2)	(3)		(7)		(4)		(9)		(5)		(8)		(6)	

A tautology

In Exercises 16–25, p has truth value T and q has truth value F.

17. $\sim$ p $\rightarrow$ q
    F   T   **T**   F

19. q $\rightarrow$ p
    F   **T**   T

21. $(p$ $\oplus$ $q)$ $\rightarrow$ p
    T   T   F   **T**   T

23. $(p$ $\wedge$ $\sim$ $q)$ $\rightarrow$ $(\sim$ p $\oplus$ $q)$
    T   T   T   F   **F**   F   T   F   F

25. p $\rightarrow$ $[p$ $\wedge$ $(p$ $\oplus$ $q)]$
    T   **T**   T   T   T   T   F

27. $p \leftrightarrow q$

29. $q \rightarrow p$; hypothesis q; conclusion: p

31. $q \rightarrow p$; hypothesis q; conclusion p

33. $\sim p \rightarrow \sim q$; hypothesis $\sim p$; conclusion $\sim q$

35. $\sim p \rightarrow \sim q$; hypothesis: $\sim p$; conclusion: $\sim q$;
    F T **T**   F T   TRUE

37. $\sim q \rightarrow \sim p$; hypothesis: $\sim q$; conclusion: $\sim p$;
    F T **T**   F T   TRUE

39. a. hyp: The plant grows.
      con: The plant is exposed to sunlight.

    b. hyp: The train stops at the station.
      con: A passenger requests the stop.

    c. hyp: People are healthy.
      con: People live a long life.

    d. hyp: I will go to the store.
      con: Jane goes to the store.

41. a. If City Sanitation collects the garbage, then the mayor calls.

    b. The price of beans goes down if there is no drought.

    c. Goldfish swim in Lake Erie if Lake Erie is fresh water.

    d. Tap water is not salted if it boils slowly.

43. a. $Z = 0 + 0 = 0$, so the condition fails and $A = 4$.

    b. $Z = 8 + (-8) = 0$, so the condition fails and $A = 4$.

    c. $Z = -3 + 3 = 0$, so the condition fails and $A = 4$.

    d. $X = -3 \not> 0$, so the condition fails and $A = 4$.

    e. $X = 8 > 0$ and $Z = 8 + (-3) = 5 \neq 0$, so the condition is met and $A = 6$.

    f. $X = 3 > 0$ and $Z = 3 + (-8) = -5 \neq 0$, so the condition is met and $A = 6$.

45. a. $B = -6 < 0$, so the condition is met and $Y = 7$.

    b. $B = -6 < 0$, so the condition is met and $Y = 7$.

    c. $C = (-2)(6) = -8 \not\geq 10$ and $B \not< 0$, so the condition fails and $Y = 0$.

    d. $B = -1 < 0$, so the condition is met and $Y = 7$.

    e. $A = 4 \not< 0$ and $B = 3 \not< 0$, so the condition fails and $Y = 0$.

    f. $A = 3 \not< 0$ and $B = 1 \not< 0$, so the condition fails and $Y = 0$.

47. a. A = −1 < 0 and B = −2 < 0, so the condition is met and C = (−1)(−2) + 4 = 6.

   b. B = 8 ≥ 6, so the condition is met and C = (−2)(8) + 4 = −12

   c. B = 3 ≮ 0 and B = 3 ≱ 6, so the condition fails and C = 0.

   d. A = 3 ≮ 0 and B = −2 ≱ 6, so the condition fails and C = 0.

   e. B = 8 ≥ 6, so the condition is met and C = (3)(8) + 4 = 28.

   f. A = 3 ≮ 0 and B = −3 ≱ 6, so the condition fails and C = 0.

49. a. C = 0 − 0 = 0, B = 0, so the condition fails and D = 0, X = 3.

   b. C = 6 − 3 = 3 ≮ 0, B = 3 ≮ 0, so the condition fails and D = 0, X = 3.

   c. C = −5 − 3 = −8 < 0, so the condition is met and D = −40, X = −37.

   d. C = 3 − 5 = −2 < 0, so the condition is met and D = −10, X = −7.

   e. B = −3 < 0, so the condition is met. C = 8 and D = 40, X = 43.

   f. B = −3 < 0, so the condition is met. C = −2 and D = 5(−2) = −10, X = −7.

Exercises 4

1.   [(*p*   →   *q*)   ∧   *q*]   →   *p*
      F   T   T   T   T   **F**   F

When *p* is false and *q* is true, the statement is FALSE.

3.   *p*   *q*   (*p*   →   *q*)   ↔   ~   (*p*   ∧   ~   *q*)

p	*q*	(*p*	→	*q*)	↔	~	(*p*	∧	~	*q*)
T	T	T	T	T	**T**	T	T	F	F	T
T	F	T	F	F	**T**	F	T	T	T	F
F	T	F	T	T	**T**	T	F	F	F	T
F	F	F	T	F	**T**	T	F	F	T	F
		(1)	(2)		(5)	(7)	(6)		(4)	(3)

5. a.   *p*   ~   *p*   ↔   *p* | *p*

p	~	*p*	↔	*p*	|	*p*
T	F	T	**T**	T	F	T
F	T	F	**T**	F	T	F
(1)	(2)		(4)		(5)	

b.

p	q	(p	∨	q)	↔	(p	|	p)	|	(q	|	q)
T	T	T	T	T	**T**	T	F	T	T	T	F	T
T	F	T	T	F	**T**	T	F	T	T	F	T	F
F	T	F	T	T	**T**	F	T	F	T	T	F	T
F	F	F	F	F	**T**	F	T	F	F	F	T	F
(1)	(2)		(5)		(7)		(3)			(6)		(4)

c.

p	q	(p	∧	q)	↔	(p	|	q)	|	(p	|	q)
T	T	T	T	T	**T**	T	F	T	T	T	F	T
T	F	T	F	F	**T**	T	T	F	F	T	T	F
F	T	F	F	T	**T**	F	T	T	F	F	T	T
F	F	F	F	F	**T**	F	T	F	F	F	T	F
(1)	(2)		(5)		(7)		(3)			(6)		(4)

d. $(p \to q) \Leftrightarrow \sim p \vee q \Leftrightarrow ((\sim p)|(\sim p))|(q|q) \Leftrightarrow \sim(\sim p)|(q|q) \Leftrightarrow p|(q|q)$

e. $p|q \Leftrightarrow \sim(\sim p)|\sim(\sim q) \Leftrightarrow (\sim p|\sim p)|(\sim q|\sim q) \Leftrightarrow (\sim p) \vee (\sim q) \Leftrightarrow \sim(p \wedge q)$

7.

p	q	c	(p	→	q)	↔	[(p	∧	~	q)	→	c)
T	T	F	T	T	T	**T**	T	F	F	T	T	F
T	F	F	T	F	F	**T**	T	T	T	F	F	F
F	T	F	F	T	T	**T**	F	F	F	T	T	F
F	F	F	F	T	F	**T**	F	F	T	F	T	F
(1)	(2)	(3)		(6)		(8)		(5)	(4)		(7)	

9. False: consider p FALSE and q TRUE.

11. $p \oplus q \Leftrightarrow (p \vee q) \wedge \sim(p \wedge q) \Leftrightarrow \sim[\sim(p \vee q) \vee \sim(\sim p \vee \sim q)]$

13. a. $\sim(p \wedge q) \Leftrightarrow \sim p \vee \sim q$
Arizona does not border California, or Arizona does not border Nevada.

b. $\sim(p \vee q) \Leftrightarrow \sim p \wedge \sim q$
There are no tickets available, and the agency cannot get tickets.

    c. $\sim(p \vee q) \Leftrightarrow \sim p \wedge \sim q$
         The killer's hat was neither white nor gray.

15. a. $\sim(p \rightarrow q) \Leftrightarrow p \wedge \sim q$
         I have a ticket to the theater, and I did not spend a lot of money.

    b. $\sim(p \rightarrow q) \Leftrightarrow p \wedge \sim q$
         Basketball is played on an indoor court, and the players do not wear sneakers.

    c. $\sim(p \rightarrow q) \Leftrightarrow q \wedge \sim q$
         The stock market is going up, and interest rates are not going down.

    d. $\sim(p \rightarrow q) \Leftrightarrow p \wedge \sim q$
         Humans have enough water and humans are not staying healthy.

17. Statement: $p \rightarrow q$
    Contrapositive: $\sim q \rightarrow \sim p$
    Converse: $q \rightarrow p$

    a. Contrapositive: If a bird is not a hummingbird, then it is not small (F).
        Converse: If a bird is a hummingbird, then it is small (T).

    b. Contrapositive: If two nonvertical lines are not parallel, they do not have the same slope. (T)
        Converse: If two nonvertical lines are parallel, they have the same slope. (T)

    c. Contrapositive: If we are not in France, then we are not in Paris (T).
        Converse: If we are in France, then we are in Paris (F).

    d. Contrapositive: If you can legally make a U-turn, then the road is not one-way (T).
        Converse: If you cannot leally make a U-turn, then the road is one-way (F).

19. Use double negation to turn the liar into a truth-teller. Instead of asking a guard, "Is your door the door to freedom?," ask either guard, "If I asked you whether your door was the door to freedom, would you say yes?"

Exercises 5

1. m = "Sue goes to the movies."
   r = "Sue reads."
   1. $m \vee r$ hyp.
   2. $\sim m$ hyp.
   3. r disj. syll. (1, 2)

3. a = "My allowance comes this week."
   p = "I pay the rent."
   b = "My bank account will be in the black."
   e = "I will be evicted."
   1. $(a \wedge p) \rightarrow b$ hyp.
   2. $\sim p \rightarrow e$ hyp.
   3. $\sim e \wedge a$ hyp.
   4. $\sim e$ subtr. (3)
   5. p mod. tollens (2, 4)
   6. a subtr. (3)
   7. b mod. ponens (5, 6, 1)

5. p = "The price of oil increases."
   a = "The OPEC countries are in agreement."
   d = "There is a U.N. debate."
   1. $p \rightarrow a$ hyp.
   2. $\sim d \rightarrow p$ hyp.
   3. $\sim a$ hyp.
   4. $\sim p$ mod. tollens (1, 3)
   5. d mod. tollens (2, 4)

7. g = "The germ is present."
   r = "The rash is present."
   f = "The fever is present."
   1. $g \rightarrow (r \wedge f)$ hyp.
   2. f hyp.
   3. $\sim r$ hyp.
   4. $\sim r \vee \sim f$ addition (3)
   5. $\sim(r \wedge f)$ DeMorgan (4)
   6. $\sim g$ mod. tollens (1, 5)

9. c = "The material is cotton."
   r = "The material is rayon."
   d = "The material can be made into a dress.'
   1. $(c \vee r) \rightarrow d$ hyp.
   2. $\sim d$ hyp.
   3. $\sim(c \vee r)$ mod. tollens (1, 2)
   4. $\sim c \wedge \sim r$ DeMorgan (3)
   5. $\sim r$ subtraction (4)

11. s = "Salaries go up."
m = "More people apply."
If s is false and m is true, then
$(s \rightarrow m) \wedge (m \vee s)$ is true but s is false.

$(s \rightarrow m) \wedge (m \vee s) \not\Rightarrow s$. The argument is
invalid.

13. y = "The balloon is yellow."
p = "The ribbon is pink."
h = "The balloon is filled with helium."
1. $y \vee p$ hyp.
2. $h \rightarrow \sim y$ hyp.
3. h hyp.
4. $\sim y$ mod. ponens (2, 3)
5. p disj. syllogism (1, 4)
The argument is valid.

15. p = "The papa bear sits."
m = "The mama bear stands."
b = "The baby bear crawls on the floor."
1. $p \rightarrow m$ hyp.
2. $m \rightarrow b$ hyp.
3. $\sim b$ hyp.
4. $\sim m$ mod. tollens (2, 3)
5. $\sim p$ mod. tollens (1, 4)
The argument is valid.

17. w = "Wheat prices are steady."
e = "Exports will increase."
s = "The GNP will be steady."
If w and s are true and e is false, then
$[w \rightarrow (e \vee s)] \wedge (w \vee s)$ is true but e is false.

$[w \rightarrow (e \vee s)] \wedge (w \wedge s) \not\Rightarrow e$. The argument
is invalid.

19. i = "Tim is industrious."
p = "Tim is in line for a promotion."
l = "Tim is thinking of leaving."
If l and p are true and i is false, then
$(i \rightarrow p) \wedge (p \vee l)$ is true but $l \rightarrow i$ is false.

$(i \rightarrow p) \wedge (p \vee l) \not\Rightarrow l \rightarrow i$.
The argument is invalid.

21. s = "Sam goes to the store."
m = "Sam needs milk."
$H_1 = s \rightarrow m$
$H_2 = \sim m$
$C = \sim s$
1. s $\sim C$
2. $s \rightarrow m$ H_1
3. m $\sim H_2$;
                          mod. ponens (1, 2)

23. n = "The newspaper reports the crime."
t = "Television reports the crime."
s = "The crime is serious."
k = "A person is killed."
$H_1 = (n \wedge t) \rightarrow s$
$H_2 = k \rightarrow n$
$H_3 = k$
$H_4 = t$
$C = s$
1. $\sim s$ $\sim C$
2. $(n \wedge t) \rightarrow s$ H_1
3. $\sim(n \wedge t)$ mod. tollens (2)
4. $\sim n \vee \sim t$ DeMorgan (3)
5. t H_4
6. $\sim n$ disj. syllogism (4, 5)
7. $k \rightarrow n$ H_2
8. $\sim k$ $\sim H_3$;
                          mod. tollens (6, 7)

Exercises 6

1. a. "1 is even or 1 is divisible by 3" is
FALSE.

b. "4 is even or 4 is divisible by 3" is
TRUE.

c. "3 is even or 3 is divisible by 3" is
TRUE.

d. "6 is even or 6 is divisible by 3" is
TRUE.

e. "5 is even or 5 is divisible by 3" is
FALSE.

3. a. $\forall x \, p(x)$

b. $\sim [\forall x \; p(x)]$

   c. $\forall x \sim p(x)$

   d.   (c) implies (b), since if nobody is taking a writing course, it follows that not everybody is.

5. Abby's statement: $\forall x \sim p(x)$, or $\sim[\exists x \, p(x)]$. This is surely false. Abby meant to say, "not all men cheat on their wives": $\sim[\forall x \, p(x)]$, or $\exists x \sim p(x)$.

7. a. $\forall x \, p(x)$

   b. $\exists x \sim p(x)$

   c. $\exists x \, p(x)$

   d. $\sim[\forall x \, p(x)]$

   e. $\forall x \sim p(x)$

   f. $\sim[\exists x \, p(x)]$

   g.   (b) and (d); (e) and (f) · (b) and (d) both say that there are some university professors who don't like poetry; (e) and (f) both say that no university professors like it.

9. a.   TRUE;
$p(4) = (4 \text{ is prime}) \rightarrow (4^2 + 1 \text{ is even})$ is TRUE, because the hypothesis is FALSE.

   b.   FALSE;
$p(2) = (\text{if } 2 \text{ is prime}) \rightarrow (2^2 + 1 \text{ is even})$ is FALSE.

11. a.   T; every x is either even or odd.

   b. $[\forall x \, p(x)] \vee [\forall x \, q(x)]$
         F   **F**   F

   c.   T; 5 is odd, hence even or odd, for instance.

   d.   F; no x is both even or odd.

   e.   F; 5 is not both even and odd, for instance.

   f. $[\exists x \, p(x)] \wedge [\exists x \, g(x)]$
         T   **T**   T

   g.   F; (4 is even) $\rightarrow$ (4 is odd) is false, for instance.

   h. $[\forall x \, p(x)] \rightarrow [\forall x \, g(x)]$
         T   **T**   T

13. a.   Not every dog has his day.

   b.   No men fight wars.

   c.   Not all mothers are married.

   d.   For some pot(s), there is no cover.

   e.   All children have pets.

   f.   Every month has 30 days.

15. a.   "The sum of any two nonnegative integers is greater than 12." FALSE: consider $x = 1$, $y = 2$. "Not every sum of two nonnegative integers is greater than 12."

   b.   "For any nonnegative integer, there is another nonnegative integer which, added to the first, makes a sum greater than 12." TRUE

   c.   "There is a nonnegative integer which, added to any other nonnegative integer, makes a sum greater than 12." TRUE (Try $x = 13$.)

   d.   "There are two nonnegative integers, the sum of which is greater than 12." TRUE (Try $x = 6$, $y = 7$.)

17. a.   FALSE: let $x = 2$, $y = 3$.

   b.   TRUE: for any x, let $y = x$.

   c.   TRUE: let $x = 1$.

   d.   FALSE: no y is divisible by every x.

   e.   TRUE: for any y, let $x = y$.

   f.   TRUE: any x divides itself.

19. a. $S \subseteq T$ translates as $\forall x[x \geq 8 \rightarrow x \leq 10]$.

b. No; consider $x = 11$.

21. $S = \{2, 4, 6, 8\}$, $T = \{1, 2, 3, 4, 6, 8\}$. So, $\forall x[x \in S \rightarrow x \in T]$.

23. The solutions to $(x - 8)(x - 3) = 0$ are 8 and 3. Only 3 is in U. The solutions to $x^2 = 9$ are -3 and 3. Only 3 is in U. So $S = T = \{3\}$.

Chapter 12 Supplementary Exercises

1. a. Statement

　b. Not a statement—not a declarative sentence.

　c. Statement

　d. Not a statement—"he" is not specified.

　e. Statement

2. a. If two lines are perpendicular, then their slopes are negative reciprocals of each other.

　b. If goldfish can live in a fishbowl, then the water is aerated.

　c. If it rains, then Jane uses her umbrella.

　d. If Sally gives Morris a treat, then he ate all his food.

3. a. Contrapositive: If the Yankees are not playing in Yankee Stadium, then they are not in New York City; converse: If the Yankees are playing in Yankee Stadium, then they are in New York City.

　b. Contrapositive: If the quake is not considered major, then the Richter scale does not indicate the earthquake is a 7; converse: If the quake is considered major, then the Richter scale indicates the earthquake is a 7.

c. Contrapositive: If a coat is not warm then it is not made of fur; converse: If a coat is warm then it is made of fur.

d. Contrapositive: If Jane is not in Moscow then she is not in Russia; converse: If Jane is in Moscow then she is in Russia.

4. a. $p = $ (two triangles are similar) and $q = $ (their sides are equal). $p \rightarrow q$ negated becomes $\sim(p \rightarrow q)$ or $p \wedge \sim q$, or "two triangles are similar but their sides are unequal."

　b. $U = \{$real numbers$\}$ and $p(x) = (x^2 = 5)$. $\exists x \, p(x)$ negated becomes $\forall x \sim q(x)$ or "For every real number x, $x^2 \neq 5$."

　c. $U = \{$positive integers$\}$, $p(n) = (n$ is even$)$, and $q(n) = (n^2$ is even$)$. $\forall n[p(n) \rightarrow q(n)]$ negated becomes $\exists n \sim [p(n) \rightarrow q(n)]$ or $\exists n[p(n) \wedge q(n)]$, or "There exists a positive integer n such that n is even but n^2 is not even."

　d. $U = \{$real numbers$\}$ and $p(x) = (x^2 + 4 = 0)$. $\exists x \, p(x)$ negated becomes $\forall x \sim p(x)$ or "For every real number x, $x^2 + 4 \neq 0$."

5. a.

p		p	$\vee$	$\sim$	p
T		T	**T**	F	T
F		F	**T**	T	F
		(1)	(3)	(2)	

Tautology

b.

	p	q		$(p$	$\rightarrow$	$q)$	$\leftrightarrow$	$(\sim$	p	$\vee$	$q)$
	T	T		T	T	T	**T**	F	T	T	T
	T	F		T	F	F	**T**	F	T	F	F
	F	T		F	T	T	**T**	T	F	T	T
	F	F		F	T	F	**T**	T	F	T	F
	(1)	(2)			(4)		(6)	(3)		(5)	

Tautology

c. Let p and q be TRUE.
$(p \wedge \sim q) \leftrightarrow \sim(\sim p \wedge q)$
T F FT **F** T FTFT
Not a tautology

d. Let p, q, and r be FALSE.
$[p \rightarrow (q \rightarrow r)] \leftrightarrow [(p \rightarrow q) \rightarrow r]$
F T F TF **F** F TF FF
Not a tautology

6. a.

p	q	r		p	$\rightarrow$	$(\sim$	q	$\vee$	$r)$
T	T	T		T	**T**	F	T	T	T
T	T	F		T	**F**	F	T	F	F
T	F	T		T	**T**	T	F	T	T
T	F	F		T	**T**	T	F	T	F
F	T	T		F	**T**	F	T	T	T
F	T	F		F	**T**	F	T	F	F
F	F	T		F	**T**	T	F	T	T
F	F	F		F	**T**	T	F	T	F
(1)	(2)	(3)			(6)	(4)		(5)	

b.

p	q	r		p	$\wedge$	$(q$	$\leftrightarrow$	$(r$	$\wedge$	$p)$
T	T	T		T	**T**	T	T	T	T	T
T	T	F		T	**F**	T	F	F	F	T
T	F	T		T	**F**	F	F	T	T	T
T	F	F		T	**T**	F	T	F	F	T
F	T	T		F	**F**	T	F	T	F	F
F	T	F		F	**F**	T	F	F	F	F
F	F	T		F	**F**	F	T	T	F	F
F	F	F		F	**F**	F	T	F	F	F
(1)	(2)	(3)			(6)		(5)		(4)	

7. a. True—a version of disjunctive syllogism

b. False—consider p false and q true.

8. a. True—contrapositive

b. False—consider p, q, and r false.

9. a. False—consider p false and q true.

b. True—modus tollens

10. a. $C = 3(4) + 5 = 17 > 0$ and $B = 5 > 0$, so the condition is met and $Z = 17$.

b. $B = 2 \not> 3$, so the condition fails and $Z = 100$.

c. $C = 3(-4) + 5 = -7 \not> 0$, so the condition fails and $Z = 100$.

d. $B = -2 \not> 3$, so the condition fails and $Z = 100$.

11. a. $C = 10$, so the condition is met and $Z = 5 \times 10 = 50$.

b. $C = 10$, so the condition is met and $Z = (5)(-5) = -25$.

c. $X = -10 \not> 0$, $C = 2 \not\ge 10$, so the condition fails and $Z = (-10) + (-5) = -15$.

d. $X = 2 > 0$, $Y = 5 > 0$, so the condition is met and $Z = 2 \times 5 = 10$.

12. a. Cannot be determined.

b. TRUE, by the contrapositive

c. TRUE

d. Cannot be determined

e. Cannot be determined

13. a. Cannot be determined

b. Cannot be determined

c. TRUE, by contraposition and DeMorgan

14. a. TRUE (given the additional assumption that at least two mathematicians exist)

b. FALSE; let $p(x)$ = "like rap music" and U = {mathematicians}. $\forall x\, p(x) \Leftrightarrow\ \sim \exists x\, p(x)$

c. TRUE

15. a.   Cannot be determined

   b.   Cannot be determined

   c.   True;
       $\exists x[\sim r(x)] \Leftrightarrow \sim \forall x[r(x)]$

16. t = "Taxes go up."
    s = "I sell the house."
    m = "I move to India."
    1. $t \rightarrow (s \wedge m)$ hyp.
    2. $\sim m$ hyp.
    3. $\sim s \vee \sim m$ addition (2)
    4. $\sim(s \wedge m)$ DeMorgan's (3)
    5. $\sim t$ mod. tollens (1, 4)

17. m = "I study mathematics."
    b = "I study business."
    p = "I can write poetry."
    1. $m \wedge b$ hyp.
    2. $b \rightarrow (\sim p \vee \sim m)$hyp.
    3. b subtraction (1)
    4. $\sim p \vee \sim m$ mod. ponens (2, 3)
    5. m subtraction (1)
    6. $\sim p$ disj. syllogism (4, 5)

18. d = "I shop for a dress."
    h = "I wear high heels."
    s = "I have a sore foot."
    1. $d \rightarrow h$ hyp.
    2. $s \rightarrow \sim h$ hyp.
    3. d hyp.
    4. h mod. ponens (1, 3)
    5. $\sim s$ mod. tollens (2, 4)

19. a = "Asters grow in the garden."
    d = "Dahlias grow in the garden."
    s = "It is spring."
    1. $a \vee d$ hyp.
    2. $s \rightarrow \sim a$ hyp.
    3. s hyp.
    4. $\sim a$ mod. ponens (2, 3)
    5. d disj. syllogism (1, 4)

20. t = "The professor gives a test."
    h = "Nancy studies hard."
    d = "Nancy has a date."
    s = "Nancy takes a shower."
    $H_1 = t \rightarrow h$
    $H_2 = d \rightarrow s$
    $H_3 = \sim t \rightarrow \sim s$
    $H_4 = d$
    $C = h$
    1. $\sim h$ $\sim C$
    2. $t \rightarrow h$ H_1
    3. $\sim t$ Mod. tollens (1, 2)
    4. $\sim t \rightarrow \sim s$ H_3
    5. $\sim s$ mod. ponens (3, 4)
    6. $d \rightarrow s$ H_2
    7. $\sim d$ $\sim H_4$; mod. tollens (5, 6)
    Contradiction

Chapter 13

Exercises 1

1. a. G_1: no loops, parallel edges from c to d
G_2: loop at e, no parallel edges
G_3: no loops or parallel edges

b. G_1:

Vertex	a	b	c	d	e	f
Degree	1	2	3	3	1	0

G_2:

Vertex	a	b	c	d	e
Degree	4	4	4	4	6

G_3:

Vertex	u	v	w	x	y	z
Degree	2	2	2	2	2	2

c. $G_1:4; G_2:0; G_3:0$

d. $G_1: 1+2+3+3+1+0 = 10 = 2(5)$
$G_2: 4+4+4+4+6 = 22 = 2(11)$
$G_3: 2+2+2+2+2+2 = 12 = 2(6)$

3.

5. Drawing a graph of this situation would result in a graph with 13 vertices of degree 3. Since the number of vertices of odd degree must be even, this is impossible.

7. a.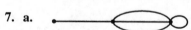

b.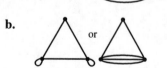

c. Not possible since the number of vertices of odd degree must be even.

9. The graph must have 10 vertices. Since there are 15 edges, the sum of the degrees of the vertices is $2(15) = 30$. Since each vertex has degree 3, there are 10 vertices.

11. a.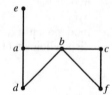

b.

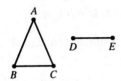

c.

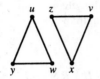

13. G_1 has 6 vertices and 5 edges while G_2 has 5 vertices and 4 edges, thus they cannot be equivalent.

15. Both graphs have 6 vertices and 5 edges.

Vertices	u	v	w	x	y	z		a	b	c	d	e	f
Degree	1	4	1	1	2	1		1	2	1	4	1	1

Pair v with d and y with b. u, w, and x can be paired with c, f, and e in any order and z is paired with a. The edges then pair up properly.

17. Not equivalent, G_1 has a vertex of degree 1(g) while G_2 has no vertices of degree one.

19. Equivalent; $A \leftrightarrow a, B \leftrightarrow b, C \leftrightarrow c, D \leftrightarrow d,$ $E \leftrightarrow e, F \leftrightarrow f, g \leftrightarrow G$

21. Equivalent. The correspondence is $A \leftrightarrow y, \ B \leftrightarrow v, \ C \leftrightarrow u, \ D \leftrightarrow x, \ E \leftrightarrow w$

23. Listing the corners in order, one possible route is $A, k, m, j, n, m, z, y, n, o, p, q, r, p, y,$ $x, r, s, t, l, u, t, x, w, z, A, B, w, u, v, B$.

25.

Exercises 2

1. a. 3

b. 5

c. 7

d. 5

e. 4

f. 7

3. a. Length 4. Not a simple path, closed path, or a circuit.

b. Length 3. A simple path but not a closed path or a circuit since the beginning and ending vertices are different.

c. Length 5. Not a simple path, closed path, or a circuit since vertex d is repeated and the beginning and ending vertices are different.

d. Length 5. A closed path and a simple circuit since it begins and ends at a and no edges or vertices are repeated.

e.   Length 4. Not a simple path, closed
path, or a circuit since vertex *f* is
repeated and the beginning and ending
vertices are different.

f.   Length 6. A closed path and a simple
circuit since it begins and ends at *a* and
no edges or vertices are repeated.

5. a.   *A B C D, A C D, A C F D, A C F E D*
(other paths are possible.)

b.   *C A B C, C F D C, C D E F C*;
yes.

7.   Only graph (a) is connected. In graph (b),
there is no path connecting any of the
vertices *u*, *v*, *w*, and *x* to any of the vertices
r, *s*, *t*, and *y*. In graph (c), there is no path
connecting vertices *A*, *B*, and *C* to any of the
vertices *D* and *E*.

9.   No, with 5 vertices and 7 edges, the graph
must be connected. Suppose one vertex is
isolated. With no loops or parallel edges,
there can be at most 6 edges between the
remaining 4 vertices. Any other
disconnection results in fewer edges.
Yes, with 5 vertices and 5 edges, there are
several possible disconnected graphs.

11.   (a) is acyclic. (b) has a circuit going from *d*
to *d*, namely the loop at *d*. c has a circuit
d e f.

13. a.   *v u w y u x y v*
(Other answers are possible.)

b.   *v r u t y x z w s z t r s v*
(Other answers are possible.)

15.

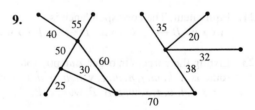

Yes, since the graph is connected and every
vertex is of even degree, it contains an Euler
circuit, which is the same as the route being
possible. One route is *a c b e f d c f a*.

17.   Yes; *A B F G C D E B C A*.

19.   Yes, it must begin on *D* or *E* and end on the
other.

21.   *b* to *d*: *b c d* or *b e d*; length: 2
d to *g*: *d e a f g*; length: 4
c to *g*: *c a f g*; length: 3

Exercises 3

1. a.   *S T Z X W V U Y R S*

b.   *b a h i g j f e d c b*

3. a.   No Euler or Hamiltonian circuits exist.

b.   Both Euler and Hamiltonian circuits
exist. (Euler: *r s t u v w x t y s x y r*,
Hamiltonian: *r s t u v w x y r*)

c.   No Euler circuits exist. Hamiltonian
circuits exist.
(Hamiltonian: *a b c d e f g h n*)

5.   No, *G* might contain loops or be
disconnected.

7.   Using Principle 2, the graph would have to
have $\frac{1}{2}(7)(6) + 2 = 23$ edges.

9.

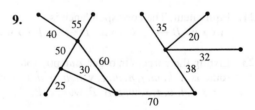

11. a.   Bipartite

b.   Not bipartite, 3 colors

c.   Not bipartite, 3 colors

13. a.   There is no way for a path to be closed
without going through a vertex
connected to the middle vertex more
than once.

b. There is no circuit that goes through both sides of the graph without going through the middle vertex more than once.

c. There is no way for a path to be closed without going through at least one vertex more than once.

15. Let each course be denoted be a vertex with an edge joining two vertices if and only if at least one student needs to take both courses.

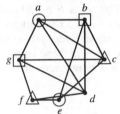

The number of time periods that are needed is the chromatic number of the graph which is 4. One coloring is shown.

Exercises 4

1. a. Neither; there is no path, from vertex u to vertex x, for instance.

b. Connected, but not strongly connected since there is no path that goes to vertex x.

c. Both connected and strongly connected.

3. a.
Vertex	A	B	C	D	E	F
id	1	2	2	1	2	0
od	1	2	1	2	0	2

b. The indegree is the number of procedures that the procedure calls, the outdegree is the number of procedures calling the procedure.

5. a. $A\,B\,A$

b. Acyclic

c. Acyclic

7. a. $p\,r\,s\,z\,y\,x\,z\,u\,v\,w\,p\,s\,w\,u\,p$ is an Euler circuit. No Euler path.

b. No Euler circuit or path since id $C = 1$ while od $C = 3$.

9. $A\,H\,G\,A\,F\,G\,E\,D\,C\,E\,B\,F\,C\,B\,A$ is an Euler circuit.

11. There are six paths from S to E. The critical path is $SA_1A_2A_3A_6A_7A_8A_{16}A_{11}E$:
time $= 2 + 2 + 8 + 3 + 4 + 3 + 2 + 1$
$= 25$ days
The shortest amount of time is 25 days.

13. The critical path is Start $ACDEF$ End.

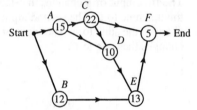

15.

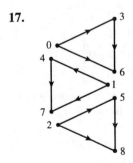

17.

19. a.

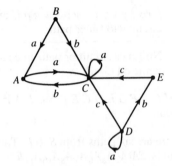

b. After the first input of b, the machine is in state A. The second input of b generates an error message.

c. The first input of b changes the state of the machine to C. The second input of b changes the state to A. The input of a changes the state to C.

Exercises 5

1. a. $A(G) = A = \begin{bmatrix} 0 & 1 & 0 & 0 & 0 \\ 1 & 0 & 1 & 1 & 0 \\ 0 & 1 & 0 & 0 & 0 \\ 0 & 1 & 0 & 0 & 1 \\ 0 & 0 & 0 & 1 & 0 \end{bmatrix}$

b. $A^2 = \begin{bmatrix} 1 & 0 & 1 & 1 & 0 \\ 0 & 3 & 0 & 0 & 1 \\ 1 & 0 & 1 & 1 & 0 \\ 1 & 0 & 1 & 2 & 0 \\ 0 & 1 & 0 & 0 & 1 \end{bmatrix}$

The 1,4 entry of A^2 is 1, so there is 1 path of length 2 from v, to v_4.

c. $M = I + A + A^2 + A^3 + A^4$

$= \begin{bmatrix} 5 & 4 & 4 & 5 & 1 \\ 4 & 14 & 4 & 5 & 5 \\ 4 & 4 & 5 & 5 & 1 \\ 5 & 5 & 5 & 9 & 3 \\ 1 & 5 & 1 & 3 & 4 \end{bmatrix}$

so the reachability matrix is

$\begin{bmatrix} 1 & 1 & 1 & 1 & 1 \\ 1 & 1 & 1 & 1 & 1 \\ 1 & 1 & 1 & 1 & 1 \\ 1 & 1 & 1 & 1 & 1 \\ 1 & 1 & 1 & 1 & 1 \end{bmatrix}.$

Also note that G is a simple connected graph.

3. a. G is not connected. In $A^2(G)$, the entries in columns 4 and 5 of rows 1, 2, and 3 will all be zeros. Similarly, the entries in rows 4 and 5 of columns 1, 2, and 3 will be zeros. Every power of A (G) will have zeros in these entries because of the way matrices are multiplied.

b.

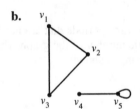

5. a. $A(G) = \begin{bmatrix} 0 & 1 & 1 & 0 & 0 \\ 1 & 0 & 1 & 0 & 0 \\ 1 & 1 & 0 & 1 & 0 \\ 0 & 0 & 1 & 0 & 0 \\ 0 & 0 & 0 & 0 & 1 \end{bmatrix}$

b. $A^3(G) = \begin{bmatrix} 2 & 3 & 4 & 1 & 0 \\ 3 & 2 & 4 & 1 & 0 \\ 4 & 4 & 2 & 3 & 0 \\ 1 & 1 & 3 & 0 & 0 \\ 0 & 0 & 0 & 0 & 1 \end{bmatrix}$

Since the 1, 4 entry of $A^3(G)$ is one, there is 1 path of length 3 from v_1.to $v_4$1

c. The reachability matrix is

$$\begin{bmatrix} 1 & 1 & 1 & 1 & 0 \\ 1 & 1 & 1 & 1 & 0 \\ 1 & 1 & 1 & 1 & 0 \\ 1 & 1 & 1 & 1 & 0 \\ 0 & 0 & 0 & 0 & 1 \end{bmatrix}$$

This can be seen from the graph.

7. a.
$$A = \begin{array}{c} \\ a \\ b \\ c \\ d \\ e \end{array}\begin{array}{c} \begin{array}{ccccc} a & b & c & d & e \end{array} \\ \begin{bmatrix} 0 & 1 & 1 & 0 & 0 \\ 1 & 0 & 1 & 0 & 0 \\ 1 & 1 & 0 & 1 & 0 \\ 0 & 0 & 1 & 0 & 1 \\ 0 & 0 & 0 & 1 & 0 \end{bmatrix} \end{array}$$

b. A^6 has no zero entries. The smallest power is 6.

c. If there is a path from vertex u to vertex v of length l, then it is easy to find a path from u to v of length $l + 2m$ by following the path of length l from u to v and then going from v to an adjacent vertex and back to v m times. Also, there are paths from a non-isolated vertex back to itself of any even length. In the circuit abc there are paths of length 1 and 2 between any two distinct vertices, hence paths of any length. In abc there are paths of length 2 and 3 from any vertex back to itself, hence paths of any length 2 or greater. There are paths of length 2 and 3 from a or b to d and paths of length 3 and 4 from a or b to e, hence paths of any length 3 or greater. There are paths from c to d of length 1, 3, and 4 and paths from c to e of length 2, 4, and 5, hence paths of any length 5 or greater. There are paths of length 1, 3, 5, and 6 between d and e, hence paths of any length 6 or greater. Since 6 is even, there are paths of length 6 from d and e back to themselves. So we are guaranteed a path of length 6 from any vertex to another. Thus A^6 has no nonzero entries.

9.
$v_1 \quad v_2 \quad v_3 \quad v_4$

11. a.
$$A = \begin{array}{c} \\ a \\ b \\ c \\ d \end{array}\begin{array}{c} \begin{array}{cccc} a & b & c & d \end{array} \\ \begin{bmatrix} 0 & 1 & 1 & 0 \\ 1 & 0 & 0 & 0 \\ 0 & 1 & 0 & 0 \\ 0 & 1 & 0 & 0 \end{bmatrix} \end{array}$$

b.
$$A^4 = \begin{array}{c} \\ a \\ b \\ c \\ d \end{array}\begin{array}{c} \begin{array}{cccc} a & b & c & d \end{array} \\ \begin{bmatrix} 1 & 2 & 1 & 0 \\ 1 & 1 & 1 & 0 \\ 1 & 1 & 0 & 0 \\ 1 & 1 & 0 & 0 \end{bmatrix} \end{array}$$

From A^4, there are 2 paths of length 4 from a to b.

c. *abacb, acbab*

d.
$$I + A + A^2 + A^3 = \begin{array}{c} \\ a \\ b \\ c \\ d \end{array}\begin{array}{c} \begin{array}{cccc} a & b & c & d \end{array} \\ \begin{bmatrix} 3 & 3 & 2 & 0 \\ 2 & 3 & 1 & 0 \\ 1 & 2 & 2 & 0 \\ 1 & 2 & 1 & 1 \end{bmatrix} \end{array}$$

$$R = \begin{array}{c} \\ a \\ b \\ c \\ d \end{array}\begin{array}{c} \begin{array}{cccc} a & b & c & d \end{array} \\ \begin{bmatrix} 1 & 1 & 1 & 0 \\ 1 & 1 & 1 & 0 \\ 1 & 1 & 1 & 0 \\ 1 & 1 & 1 & 1 \end{bmatrix} \end{array}$$

e. D is connected, but not strongly connected since there are zeros in the reachability matrix of D. For instance, there is no path to d from any other vertex.

13. a.

Vertex	v_1	v_2	v_3	v_4	v_5
id	3	2	2	3	1
od	2	2	2	3	2

The indegree of v_i is the sum of the entries in the column of A corresponding to v_i, the outdegree of

v_i is the sum of the entries in the row of A corresponding to v_i.

b.

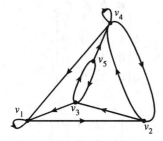

c. $A^3 = \begin{bmatrix} 3 & 2 & 1 & 2 & 1 \\ 3 & 3 & 2 & 3 & 0 \\ 3 & 2 & 1 & 2 & 1 \\ 5 & 4 & 2 & 4 & 1 \\ 3 & 3 & 2 & 3 & 0 \end{bmatrix}$

From A^3, there are 2 paths from v_1 to v_4.

d. $v_1 v_2 v_4 v_4,\ v_1 v_1 v_2 v_4$

Exercises 6

1.

3.

5. a. d, e, i, j, k, h

    b. c, d, e, f, g, h

c. $x, 6, y$

7. a. a

    b. b

    c. $*$

9. Not possible since a tree with 7 vertices must have 6 edges.

11. Not possible, a connected graph with 4 vertices and 3 edges must be a tree.

13. Not possible since a tree with 3 vertices must have 2 edges.

15.

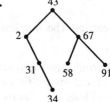

17. $[a \div (b + c)] - (b \wedge c)$

19.

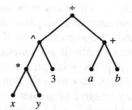

21. 7 questions, 10 questions

Each question reduces the set of numbers being searched by half. Thinking of going backwards through the search, at each step the size of the set of numbers is being doubled. Thus, the number of questions will be a power of 2. One question suffices if the original set contains $2 = 2^1$ numbers. Two questions will be enough if the original set contains $4 = 2^2$ or fewer numbers. Thus, the maximum number of questions is the smallest exponent n such that 2^n is greater than the size of the set being searched. Since $2^6 < 100 < 2^7$ and $2^9 < 1000 < 2^{10}$, it takes a maximum of 7 questions if the number is between 1 and 100; 10 questions if the number is between 1 and 1000.

Chapter 13 Supplementary Exercises

1. G_1:

Vertex	a	b	c	d
Degree	2	3	2	5

sum of degrees = 12; number of edges = 6

G_2:

Vertex	r	s	t	u	v	w	x
Degree	3	2	3	3	4	4	3

sum of degrees = 22;
number of edges = 11

2. a. G_1: *abc*; G_2: *uvwrst*

   b. G_1: *abcdba*; G_2: *trstvwrt*

   c. G_1: *dbcdd*; G_2: *rtvuwvxwr*

3. G_1: weight = 9

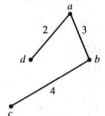

G_2: weight = 15

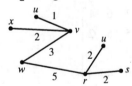

4. Neither graph has an Euler circuit since both have vertices of odd degree.

5. G_1:

Vertex	a	b	c	d	e	f	g	h
Degree	2	3	3	4	2	3	5	4

There are 4 vertices of odd degree.

G_2:

Vertex	a	b	c	d	e	f	g	h
Degree	2	4	2	4	4	2	4	4

There are 0 vertices of odd degree.

6. a. G_1: $a\,h\,c\,d\,e$
G_2: $g\,f\,d\,e\,b$

b. G_1: $f\,d\,c\,g\,d\,e\,f$
G_2: $b\,e\,h\,g\,e\,d\,c\,b$

c. G_1: $a\,b\,g\,c\,h\,b\,a$
G_2: $a\,b\,h\,e\,b\,a$

d. G_1: $a\,h\,g\,c\,h\,b\,a$
G_2: $a\,b\,h\,e\,g\,h\,a$

7. G_1: weight = 19

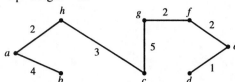

G_2: weight = 37

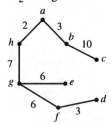

8. G_1 does not have an Euler circuit since it has vertices of odd degree.
In G_2, $a\,b\,c\,d\,e\,b\,h\,e\,g\,d\,f\,g\,h\,a$ is an Euler circuit.

9. G_1:

$$A = \begin{bmatrix} 0 & 1 & 0 & 1 & 1 \\ 1 & 0 & 1 & 1 & 1 \\ 0 & 1 & 0 & 1 & 0 \\ 1 & 1 & 1 & 0 & 1 \\ 1 & 1 & 0 & 1 & 0 \end{bmatrix}$$

$$A^3 = \begin{bmatrix} 6 & 9 & 4 & 9 & 7 \\ 9 & 8 & 7 & 9 & 9 \\ 4 & 7 & 2 & 7 & 4 \\ 9 & 9 & 7 & 8 & 9 \\ 7 & 9 & 4 & 9 & 6 \end{bmatrix}$$

From A^3, there are 9 paths of length 3 from v_1 to v_4.

G_2:

$$A = \begin{bmatrix} 0 & 1 & 0 & 1 & 0 \\ 1 & 0 & 1 & 1 & 0 \\ 0 & 1 & 0 & 1 & 0 \\ 1 & 1 & 1 & 0 & 1 \\ 0 & 0 & 0 & 1 & 0 \end{bmatrix}$$

$$A^3 = \begin{bmatrix} 2 & 5 & 2 & 6 & 1 \\ 5 & 4 & 5 & 6 & 2 \\ 2 & 5 & 2 & 6 & 1 \\ 6 & 6 & 6 & 4 & 4 \\ 1 & 2 & 1 & 4 & 0 \end{bmatrix}$$

From A^3, there are 6 paths of length 3 from v_1 to v_4.

10.   Only G_2 is equivalent to G_1. G_3 has six vertices while G_1 has five and G_4 has five edges while G_1 has six.

11.

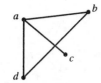

G' is equivalent to G. Associate u to b, v to d, w to a, and x to c.

12. The path does not exist since rooms B, D, and H have only one doorway.
The existence of the path is equivalent to the existence of an Euler circuit in the associated graph. There is no Euler circuit in the graph since the vertices corresponding to rooms B, D, E, and H will have odd degree.

13. 3 colors are needed since there are two circuits of length 3.

14. Start A B D E F End

15.

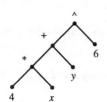

Since 127 succeeds 120, proceed to the right child of 120. Then, since 127 precedes 187, proceed to the left child of 187. 127 precedes 134, so we add it as the left child of 134.

16. a.

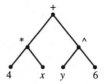

b.

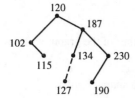

c.

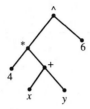

17.

Path	Time
Start *A B E G* End	$20 + 60 + 30 + 20 = 130$
Start *A B C G* End	$20 + 60 + 50 + 20 = 150$
Start *A B D F G* End	$20 + 60 + 40 + 20 + 20 = 160$
Start *A D F G* End	$20 + 40 + 20 + 20 = 100$

The critical path is Start *A B D F G* End.

18. The family needs 3 cars. The number of cars needed is the chromatic number of the graph, with each color representing a car. Three colors are needed to color the graph.

19. Construct a graph with each vertex representing a country and each edge a shared border.

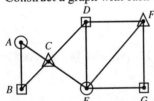

A coloring using 3 colors is shown. Since the graph contains circuits of length 3, this is the minimal number of colors needed.

20. The tournament is not transitive since *A B D A* is a circuit. One possible ranking is *D E A B C*. This is not unique.

YOU SHOULD CAREFULLY READ THE FOLLOWING TERMS AND CONDITIONS BEFORE USING THIS DISKETTE PACKAGE. USING THIS DISKETTE PACKAGE INDICATES YOUR ACCEPTANCE OF THESE TERMS AND CONDITIONS.

Prentice-Hall, Inc. provides this program and licenses its use. You assume responsibility for the selection of the program to achieve your intended results, and for the installation, use, and results obtained from the program.

LICENSE GRANT

You hereby accept a nonexclusive, nontransferable, permanent license to install and use the program ON A SINGLE COMPUTER at any given time. You may copy the program solely for backup or archival purposes in support of your use of the program on the single computer. You may not modify, translate, disassemble, decompile, or reverse engineer the program, in whole or in part.

LIMITED WARRANTY

THE PROGRAM IS PROVIDED "AS IS" WITHOUT WARRANTY OF ANY KIND, EITHER EXPRESSED OR IMPLIED, INCLUDING, BUT NOT LIMITED TO, THE IMPLIED WARRANTIES OF MERCHANTABILITY AND FITNESS FOR A PARTICULAR PURPOSE. THE ENTIRE RISK AS TO THE QUALITY AND PERFORMANCE OF THE PROGRAM IS WITH YOU. SHOULD THE PROGRAM PROVE DEFECTIVE, YOU (AND NOT PRENTICE HALL, INC. OR ANY AUTHORIZED DISTRIBUTOR) ASSUME THE ENTIRE COST OF ALL NECESSARY SERVICING, REPAIR, OR CORRECTION. NO ORAL OR WRITTEN INFORMATION OR ADVICE GIVEN BY PRENTICE-HALL, INC., ITS DEALERS, DISTRIBUTORS, OR AGENTS SHALL CREATE A WARRANTY OR INCREASE THE SCOPE OF THIS WARRANTY.

SOME STATES DO NOT ALLOW THE EXCLUSION OF IMPLIED WARRANTIES, SO THE ABOVE EXCLUSION MAY NOT APPLY TO YOU. THIS WARRANTY GIVES YOU SPECIFIC LEGAL RIGHTS AND YOU MAY ALSO HAVE OTHER RIGHTS THAT VARY FROM STATE TO STATE.

Prentice-Hall, Inc. does not warrant that the function contained in the program will meet your requirements or that the operation of the program will be uninterrupted or error free.

However, Prentice-Hall, Inc. warrants the diskette(s) on which the program is furnished to be free from defects in materials and workmanship under normal use for a period of ninety (90) days from the date of delivery to you as evidenced by a copy of your receipt.

The program should not be relied on as the sole basis to solve a problem whose incorrect solution could result in injury to person or property. If the program is employed in such a manner, it is at the user's own risk and Prentice-Hall, Inc. explicitly disclaims all liability for such misuse.

LIMITATION OF REMEDIES

Prentice-Hall's entire liability and your exclusive remedy shall be:

1.   the replacement of any diskette not meeting Prentice-Hall, Inc.'s "Limited Warranty" and that is returned to Prentice-Hall, or

2.   if Prentice-Hall is unable to deliver a replacement diskette or cassette that is free of defects in materials or workmanship, you may terminate this Agreement by returning the program.

IN NO EVENT WILL PRENTICE-HALL BE LIABLE TO YOU FOR ANY DAMAGES, INCLUDING ANY LOST PROFITS, LOST SAVINGS, OR OTHER INCIDENTAL OR CONSEQUENTIAL DAMAGES ARISING OUT OF THE USE OR INABILITY TO USE SUCH PROGRAM EVEN IF PRENTICE-HALL OR AN AUTHORIZED DISTRIBUTOR HAS BEEN ADVISED OF THE POSSIBILITY OF SUCH DAMAGES, OR FOR ANY CLAIM BY ANY OTHER PARTY.

SOME STATES DO NOT ALLOW THE LIMITATION OR EXCLUSION OF LIABILITY FOR INCIDENTAL OR CONSEQUENTIAL DAMAGES, SO THE ABOVE LIMITATION MAY NOT APPLY TO YOU.

GENERAL

You may not sublicense, assign, or transfer the license or the program except as expressly provided in this Agreement. Any attempt otherwise to sublicense, assign, or transfer any of the rights, duties, or obligations hereunder is void.

This Agreement will be governed by the laws of the State of New York.

Should you have any questions concerning this Agreement, you may contact Prentice-Hall, Inc. by writing to:

Prentice Hall
Engineering, Science, and Math Division
One Lake Street
Upper Saddle River, NJ 07458

YOU ACKNOWLEDGE THAT YOU HAVE READ THIS AGREEMENT, UNDERSTAND IT, AND AGREE TO BE BOUND BY ITS TERMS AND CONDITIONS. YOU FURTHER AGREE THAT IT IS THE COMPLETE AND EXCLUSIVE STATEMENT OF THE AGREEMENT BETWEEN US THAT SUPERSEDES ANY PROPOSAL OR PRIOR AGREEMENT, ORAL OR WRITTEN, AND ANY OTHER COMMUNICATIONS BETWEEN US RELATING TO THE SUBJECT MATTER OF THIS AGREEMENT.

NOTICE TO GOVERNMENT END USERS
The program is provided with RESTRICTED RIGHTS. Use, duplication or disclosure by the government is subject to restrictions set forth in subdivion (b)(3)(iii) of The Rights in Technical Data and Computer Software clause 252.227-7013.

ISBN 0-13-747684-1